The Adventures of Diode Dude

Written & Illustrated by B. K. Hixson

© 2003 • B. K. Hixson • Loose in the Lab

The Adventures of Diode Dude

Copyright © 2003
First Printing • September 2003
B. K. Hixson

Published by Loose in the Lab, Inc.
9462 South 560 West
Sandy, Utah 84070

www.looseinthelab.com

Library of Congress Cataloging-in-Publication Data:

Hixson, B. K.
    The Adventures of Diode Dude/B. K. Hixson
    p. cm.-(Loose in the Lab Science Series)

    Includes index
    ISBN 1-931801-08-8
    1. Electronics experiments–juvenile literature. [1. Electronics experiments 2. Experiments] I. B. K. Hixson II. Loose in the Lab III. Title IV. Series
QP37.H24   2003
612

Printed in the United States of America
Long Live the Resistance.

Dedication

P. K. Hixson

    For my #1 Dude. May your circuits always be charged, your resistance be efficiently necessary, and your electrons be ready to zip around in gleeful joy over the myriad options that they have to explore.

Acknowledgments

I built my first crystal radio for a Scout project when I was a peanuthead and have tinkered with projects off and on. I would have to say that, despite my various traveling, readings, and burnt fingers, I learned the most hanging out with my college friend, Tim Hutson, whose electronics education was self-taught and near legendary. Driven by capitalism, he created circuits that drove lights so that the dances he put on in school would be more than loud music. I have many fond memories of stringing lights, replacing blown speakers, and dancing until they kicked us out of the cafeteria.

As for my educational outlook, the hands-on perspective, and the use of humor in the classroom, Dr. Fox, my senior professor at Oregon State University, gets the credit for shaping my educational philosophy while simultaneously recognizing that even at the collegiate level, we were onto something a little different. He did his very best to encourage, nurture, and support me while I was getting basketloads of opposition for being willing to swim upstream. There were also several colleagues who helped to channel my enthusiasm during those early, formative years of teaching: Dick Bishop, Dick Hinton, Dee Strange, Connie Ridgway, and Linda Zimmermann. Thanks for your patience, friendship, and support.

Next up are all the folks who get to do the dirty work that makes the final publication look so polished but very rarely get the credit they deserve. Our resident graphics guru, Kris Barton, gets a nod for scanning and cleaning the artwork you find on these pages, as well as putting together the graphics that make up the cover. Once we have a finished product, it has to be printed so that Gary Facente, Louisa Walker, Lisa Lachance, and the Delta Education gang can market and ship the books and collect the money.

Mom and Dad, as always, get the end credits. Thanks for the education, encouragement, and love. And for Kathy and the kids—Porter, Shelby, Courtney, and Aubrey—hugs and kisses.

Repro Rights

There is very little about this book that is truly formal, but at the insistence of our wise and esteemed counsel, let us declare: *No part of this book may be reproduced or utilized in any form or by any means, electronic or mechanical, including photocopying, recording, or by any information storage and retrieval system, without permission in writing from the publisher.* That's us.

More Legal Stuff

Official disclaimer for you aspiring scientists and lab groupies: This is a hands-on science book. By the very intent of its design, you will be directed to use common, nontoxic household items in a safe and responsible manner to avoid injury to yourself and others who are present while you are pursuing your quest for knowledge and enlightenment in the world of electronics. Just make sure that you have a fire blanket handy and a wall-mounted video camera to corroborate your story.

If, for some reason, perhaps even beyond your own control, you have an affinity for disaster, we wish you well. *But we in no way take any responsibility for any injury that is incurred to any person using the information provided in this book or for any damage to personal property or effects that is directly or indirectly a result of the suggested activities contained herein.* Translation: You're on your own, despite the fact that many have preceded you in the lab.

Less Formal Legal Stuff

If you happen to be a home schooler or very enthusiastic school teacher, please feel free to make copies of this book for your classroom or personal family use—one copy per student, up to 35 students. If you would like to use an experiment from this book for a presentation to your faculty or school district, we would be happy to oblige. Just give us a whistle, and we will send you a release for the particular lab activity you wish to use. Please contact us at the address below. Thanks.

Special Requests
Loose in the Lab, Inc.
9462 South 560 West
Sandy, Utah 84070

Table of Contents

Dedication 3
Acknowledgments 4
Repro Rights 5
Who Are You? And ... How to Use This Book 8
Lab Safety 14

    The Components 17
    Bread Board 24

The Projects
    1 *Pocket Engine* 26
    2 *Impromtu Cop Car* 29
    3 *Theater Lighting* 32
    4 *Jingle Lights* 35
    5 *Burglar Alarm* 38
    6 *Musical Line-Up* 42
    7 *Finger Switch* 45
    8 *Auditory Circuit* 48
    9 *Photo Switch* 51
    10 *Light Organ* 54
    11 *Timing With Capacitance* 57
    12 *Piezo Candle* 61
    13 *Conductivity Tester* 64
    14 *Electronic Rooster* 67
    15 *Electron Hand Jive* 70
    16 *Morse's Code Machine* 73
    17 *Simple Telephone* 76
    18 *Classic Crystal Radio* 79
    19 *Transistor Radio* 82
    20 *Flip-Flop Circuit* 86

Science Fair Projects   90

A Step-by-Step Guide: From Idea to Presentation
    Step #1: The Hypothesis   *98*
    Step #2: Gather Information   *105*
    Step #3: Design Your Experiment   *110*
    Step #4: Conduct the Experiment   *115*
    Step #5: Collect and Display Data   *117*
    Step #6: Present Your Ideas   *121*

Index   *125*
More Science Books   *128*

Who Are You? And ...

First of all, we may have an emergency at hand and we'll both want to cut to the chase and get the patient into the cardiac unit, if necessary. So, before we go too much further, **define yourself**. Please check one and only one choice listed below and then immediately follow the directions that follow *in italics*. Thank you in advance for your cooperation.

I am holding this book because ...

___ **A. I am a responsible, but panicked, parent.** My son/daughter/triplets (circle one) just informed me that his/her/their science fair project is due tomorrow. This is the only therapy I could afford on such short notice. This means that, if I were not holding this book, my hands would be encircling the soon-to-be-worm-bait's neck.

Directions: Can't say this is the first or the last time we heard that one. Hang in there, we can do this.

1. Quickly read the Table of Contents with the worm bait. Obviously, the kid is not passionate about science, or you would not be in this situation. See if you can find an idea that causes some portion of an eyelid or facial muscle to twitch.

2. Take the materials list from the lab write-up and from page 111 of the Science Fair Project section and go shopping.

3. Assemble the materials and perform the lab at least once. Gather as much data as you can.

4. Start on Step 1 of Preparing Your Science Fair Project. With any luck, you can dodge an academic disaster.

How to Use This Book

___ **B. I am worm bait.** My science fair project is due tomorrow, and there is not anything moldy in the fridge. I need a big Band-Aid in a hurry.

Directions: Same as Option A. You can decide if and when you want to clue your folks in on your current dilemma.

___ **C. I am the parent of a student who informed me that he/she has been assigned a science fair project due in six to eight weeks.** My son/daughter has expressed an interest in science books with humorous illustrations that attempt to explain electronics.

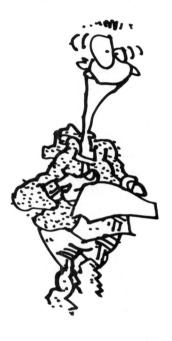

Who Are You? And ...

Directions: Well, you came to the right place. Give your kid these directions and stand back.

1. The first step is to read through the Table of Contents and see if anything grabs your interest. Read through several experiments, see if the science teacher has any of the more difficult-to-acquire materials, and ask if they can be borrowed. Play with the experiments and see which one really tickles your fancy.

2. You have plenty of time, so you can fiddle and fool with the original experiment and its derivations several times. Work until you have an original question you want to answer, and then start the process. You are well on your way to an excellent grade.

___ **D. I am a responsible student and have been assigned a science fair project due in six to eight weeks.** I am interested in electronics, and despite demonstrating maturity and wisdom well beyond the scope of my peers, I too still have a sense of humor. Enlighten and entertain me.

Directions: Cool. Being teachers, we have heard reports of this kind of thing happening, but usually in an obscure and hard-to-locate town, several states removed. Nonetheless, congratulations.

Same as Option C. You have plenty of time and should be able to score very well. We'll keep our eyes peeled when the Nobel Prizes are announced in a couple of decades.

How to Use This Book

___ **E. I am a parent who home schools my child/children.** I am always on the lookout for high-quality curriculum materials that are not only educationally sound but also kid- and teacher-friendly. I am not particularly strong in science, but I realize it is a very important topic. How is this book going to help me out?

Directions: In a lot of ways, we created this book specifically for home schoolers.

1. We have taken the National Content Standards, the guidelines that are used by all public and private schools nationwide to establish their curriculum base, and have used them as references in developing our labs.

2. When your children are done studying this unit on electronics, you want them not only to understand and explain each of the big ideas covered in the labs listed in this book, but also to be able to defend and argue their position based on experiential evidence that they have collected.

3. Building on the central concepts, we have collected and rewritten 20 hands-on science labs. Each one has been specifically selected so that it supports hands-on learning. As the kids do the science experiment, they see, smell, touch, and hear the experiment. They will store that information in several places in their brains.

Who Are You? And ...

    *For example: I can show you a recipe in a book for chocolate chip cookies and ask you to reiterate it. Or I can turn you loose in a kitchen, have you mix the ingredients, grease the pan, plop the dough on the cookie sheet, slide everything into the oven, and wait impatiently until they pop out eight minutes later. Chances are that the description given by the person who actually made the cookies is going to be much clearer because it is based on a true understanding of the process, **because it is based on experience.***

    4. *Once you have completed the experiment, you'll be able to extend the lab and vary it.*

    5. *A word about humor. Science is not usually known for being funny, even though* Bill Nye, The Science Guy, *Beaker from* Sesame Street, *and* Beakman's World *do their best to mingle the two. That's all fine and dandy, but we want you to know that we incorporate humor because it is scientifically (and educationally) sound to do so. Plus it's really at the root of our personalities. Here's what we know:*
    When we laugh ...
    a. Our pupils dilate, increasing the amount of light entering the eye.
    b. Our heart rate increases, which pumps more blood to the brain.
    c. Oxygen-rich blood to the brain means the brain is able to collect, process, and store more information. Big I.E.: increased comprehension.
    d. Laughter relaxes muscles, which can be involuntarily tense if a student is uncomfortable or fearful of an academic topic.
    e. Laughter stimulates the immune system, which will ultimately translate into overall health and fewer kids who say they are sick of science.
    f. Socially, it provides an acceptable pause in the academic routine, which then gives the student time to regroup and prepare to address some of the more difficult ideas with a renewed spirit. They can study longer and focus on ideas more efficiently.
    g. Laughter releases chemicals in the brain that are associated with pleasure and joy.

    6. *If you follow the book in the order in which it is written, you will be able to build ideas and concepts in a logical and sequential pattern. But that is by no means necessary. For a complete set of guidelines on our ideas on how to teach home-schooled kids science, check out our book,* Why's the Cat on Fire? How to Excel at Teaching Science to Your Home-Schooled Kids.

How to Use This Book

___ F. **I am a public/private school teacher,** and this looks like an interesting book to add ideas to my classroom lesson plans.

Directions: It is, and please feel free to do so.

___ G. **My son/daughter/ grandson/niece/father-in-law** is interested in science, and this looks like fun.

Directions: Congratulations on your selection. Hook them up with a pass to the local science museum and you've got the perfect Saturday afternoon gig.

___ H. **I have this incredible temptation to stick my finger into a wall socket and become one with the electron. Any ideas?**

Directions: Just one. We would like a sample of your DNA so we can start looking for the stupid gene.

Lab Safety

Contained herein are 20 lab projects to help you better understand the nature and characteristics of electronics as we currently understand these things. However, because you are on your own in this journey, we thought it prudent to share some basic wisdom and experience in the safety department.

Read the Instructions

An interesting concept, especially if you are a teenager. Take a minute before you jump in and get going to read all of the instructions as well as warnings. If you do not understand something, stop and ask an adult for help.

Clean Up All Messes

Keep your lab area clean. It will make it easier to put everything away at the end and may also prevent contamination and the subsequent germination of a species of mutant tomato bug larva. You will also find that chemicals perform with more predictability if they are not poisoned with foreign molecules.

Organize

Translation: Put it back where you get it. If you need any more clarification, there is an opening at the landfill for you.

Dispose of Poisons Properly

This will not be much of a problem with the labs that are suggested in this book. However, if you happen to wander over into one of the many disciplines that incorporates the use of advanced chemicals, then we would suggest that you use great caution with the materials, and definitely dispose of any and all poisons properly.

Practice Good Fire Safety

If there is a fire in the room, notify an adult immediately. If an adult is not in the room and the fire is manageable, smother the outbreak with a fire blanket or use a fire extinguisher. When the fire is contained, immediately send someone to find an adult. If, for any reason, you happen to catch on fire, **REMEMBER: Stop, Drop, and Roll.** Never run; it adds oxygen to the fire, making it burn faster, and it also scares the bat guano out of the neighbors when they see the neighbor kids running down the block doing an imitation of a campfire marshmallow without the stick.

Protect Your Skin

It is a good idea to always wear protective gloves whenever you are working with chemicals. This particular book does not suggest or incorporate hazardous chemicals into its lab activities. However, if you do happen to spill a chemical on your skin, notify an adult immediately and then flush the area with water for 15 minutes. It's unlikely, but if irritation develops, have your parents or another responsible adult look at it. If it appears to be of concern, contact a physician. Report any information that you have about the chemical to the doctor.

Lab Safety

Save Your Nose Hairs

Sounds like a cause celebre L.A. style, but it is really good advice. To smell a chemical to identify it, hold the open container six to ten inches down and away from your nose. Make a clockwise circular motion with your hand over the opening of the container, "wafting" some of the fumes toward your nose. This will allow you to safely smell some of the fumes, without exposing yourself to a large dose of anything noxious. This technique may help prevent a nosebleed or your lungs from accidentally getting burned by chemicals.

Wear Goggles If Appropriate

If the lab asks you to heat or mix chemicals, be sure to wear protective eyewear. Also have an eyewash station or running water available. You never know when something is going to splatter, splash, or react unexpectedly. It is better to look like a nerd and be prepared than to schedule a trip down to pick out a Seeing Eye™ dog. If you do happen to accidentally get chemicals into your eye, flush the area for 15 minutes. If any irritation or pain develops, immediately go see a doctor.

Lose the Comedy Routine

You should have plenty of time scheduled during your day to mess around, but science lab is not one of them. Horseplay breaks glassware, spills chemicals, and creates unnecessary messes—things that parents do not appreciate. Trust us on this one.

No Eating

Do not eat while performing a lab. Putting your food in the lab area contaminates your food and the experiment. This makes for bad science and worse indigestion. Avoid poisoning yourself and goobering up your labware by observing this rule.

Happy and safe experimenting!

The Components

The Cast

First things first. We are going to introduce you to some of the basic pieces of an electronics circuit. There are a lots of components out there, but because we are starting with projects that are designed to introduce you to some basic ideas and demonstrate some principles and characteristics, we start with the most common.

In this section, we will describe all of the components that you are going to use to make your projects. There is a cartoon of what each component looks like and a symbol that is used in the schematics, which show you how to build each project. So, without further ado, this is the cast of characters that will provide your education.

Capacitor

The first parts that you are going to install are called capacitors. They are storage tanks that can be filled with electricity and then emptied when the time is right. They come in a couple thousand different sizes and shapes and are used primarily as a place in a circuit where electricity can be accumulated, or stored, at a higher level until it is needed.

Capacitors are also very helpful in smoothing out any variations in the flow of power through the circuit. All capacitors have numbers printed on their sides that indicate how much electricity they will hold. They're measured in units called "Farads," which are named in honor of a famous scientist named Michael Faraday, who made a lot of discoveries in the field of electricity.

The Components

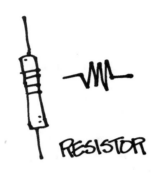

Resistor

Resistors are components that reduce (or resist) the flow of electricity. Even in materials that conduct electricity, there is always some resistance to the movement of electricity through them. Resistance is like running into a headwind: It makes it hard to move the electricity through the circuit. Most of the time, resistance is a problem to overcome because it means that electricity is wasted.

However, resistors are made to have a very specific level of resistance. Because they control the amount of electricity traveling through a circuit, they can be used to control how much electricity flows at any given time, which is very important. Some of the components you will use for these projects will need only a very small amount of electricity to do their job. The nine-volt batteries produce more electricity than they need, and without resistors, they would get overrun by a thundering herd of electrons and would burn out. The electricity straight from the battery would make some circuits run too fast to be usable, but resistors slow down the flow of electrons to a usable level.

If you look carefully, you'll see that each resistor has colored stripes. The stripes indicate how much resistance each component provides in the circuit. Each color indicates a different number, except for a gold or silver one on the end. The stripes are pretty small, so you should try to work with a good light and look closely so that you can tell the difference between a brown stripe and a purple stripe, for example. You will need six resistors to complete the projects that we have outlined in this book.

Transistor

Transistors are usually two small black components with three leads sticking out the bottom. There are only two transistors used in the projects that we have included in this book, but you must use care in handling them because they break very easily.

Transistors work like small electricity faucets. The top leg is where electricity goes in, and the bottom leg is where electricity goes out. The middle leg turns the flow on and off. If just a little bit of electricity goes into the middle leg, it will open the transistor and let a lot of electricity flow from the top leg to the bottom leg. If the middle leg doesn't get any electricity, then no electricity goes through.

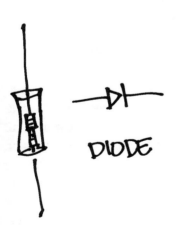

Diode

Ah, our namesake. The diode is a little glass tube in the shape of a cylinder, with wire leads coming out of both ends. A diode is a "one-way" component, like a valve or street. It is designed to let electricity pass through in one direction but not the other.

Diodes are used to guide the flow of electricity along the path in the proper direction.

The Components

Light-Emitting Diode

The bright red and green "lights" that look like the one in the picture at the right are called light-emitting diodes, or LEDs for short. Just like the diodes you learned about on the previous page, these allow electricity to flow in only one direction. In addition to controlling the flow of electricity, they also light up as the electricity passes through them.

If you look at an LED carefully, you will notice that one side of the little edge around the bottom is flat. When you put your LEDs into the circuit, you have to pay special attention to their orientation. If you put them in backwards, you will plug up your entire circuit, and no electricity will be able to flow through it.

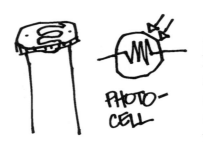

Photocell

The photocell is an orange disc with tiny stripes. This is actually a special kind of resistor that varies how much resistance it provides, depending on the amount of light that is shining on it. The darker it gets in the room where you're working, the less electricity the photocell allows to pass through.

Transformer

The funny-looking thing with five wire leads coming out of it, three on one side and two on the other, is called a transformer. A transformer, despite the fact that you may currently be confusing this component with a tiny robot, is a device that changes the voltage of the electricity passing through it to make it more appropriate for the needs of the circuit.

Piezo Transducer

A piezo transducer, pronounced **(pee-ay-zoe)**, is a round metal disc with two leads coming from the bottom. This gadget makes noise when electricity passes through it. Place the two leads on the piezo transducer through the holes in the bread board when you are installing the project. (See page 24.) The

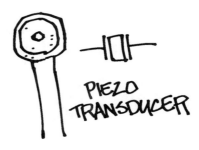

square, clear box on the top of the disc is a variable capacitor. As the name implies, it is a capacitor that can be changed or varied, which can be a very handy thing. On top of the box is a short metal tube with screw threads on the inside. This is where you adjust the capacitor to be able to fill it to the level that you want for the project.

The Components

Antenna

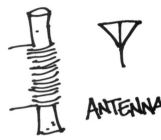

Your antenna will probably come as a black metal rod and coil of copper wire. Your antenna will pick up faint radio signals that you will actually be able to hear when you pump them through an electrical circuit attached to an earphone.

To assemble your antenna, slide the coil of wire over the metal rod, as shown. You may have to bend the coil to match the shape of the metal rod.

Earphone

The earphone looks like an ... earphone. This is the one piece of electronics equipment that should be very familiar to you. Your earphone has a very small speaker inside it and can change electricity signals into audible sound. You will attach the earphone to the bread board via the wire that comes from the bottom.

Battery

And last, but certainly not least, is the 9V battery and the battery holder. When you look at the battery holder, you will notice that there are two leads coming from the plastic cap. One is red, and the other is black. These colors indicate the direction in which electricity flows out of the battery and into the circuit.

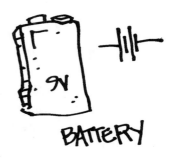

When you are installing the earphone, you will definitely want to double-check this because, if you put it in backwards, you will be without electricity.

When you are done with a circuit, always disconnect the battery and store it in a safe place.

Shopping List

Now that we have gone through the different components, here is a complete shopping list to take to the local electronics store so that you can have everything that you need to complete the projects.

1 Capacitor, .001µF	1 Transformer
1 Capacitor, .01µF	1 Variable capacitor
1 Capacitor, .05µF	2 Transistors
1 Capacitor, 4.7µF	1 Battery, 9V
1 Capacitor, 470µF	1 Diode
	2 LEDs
1 Resistor, 510Ω	1 Earphone
1 Resistor, 1KΩ	1 Piezo transducer
1 Resistor, 5.1KΩ	2 Yellow wires,
1 Resistor, 10KΩ	each 4 inches long
1 Resistor, 100KΩ	1 Photocell
1 Resistor, 470KΩ	1 Bread board
1 Resistor, 10KΩ	25 Connecting wires,
	each 3 inches long

Bread Boards

How To Use A Bread Board

A bread board is an excellent and convenient method of building circuits. Each row has a series of holes, or sockets, in it. Each socket is large enough for only one wire to fit in. All of the sockets in a row are interconnected. This makes it simple to connect components together. For example, if you place a lead of a capacitor into the socket in one row and place a lead of a resistor into a second socket in the same row, the two pieces will be connected. Continue this type of connecting, to join all of the elements in the circuit. No two rows are connected. The only way to connect two different rows is to place a wire jumper from a socket in one row to a socket in a different row. It is frequently easier to insert all of the leads of the components into separate rows and then use wires to connect them.

These are available at your local electronics shop, such as Radio Shack. They come in various sizes and shapes. A small, simple one will do the trick.

The Projects

Pocket Engine

The Experiment

This particular circuit design produces a realistic engine sound, similar to what you would hear in an arcade game. Because it is so small, we figured that we would call it your "pocket engine." If your mouth ever gets tired of creating the sound of passing cars, in the Doppler mode, we figure you can whip out your pocket engine and have at it.

Materials

Capacitor, 4.7µF
Transformer
Transistor
Diode
Connecting wires,
    each 3 inches long

Resistor, 5.1KΩ
Piezo transducer
Battery, 9V
Earphone

Schematic

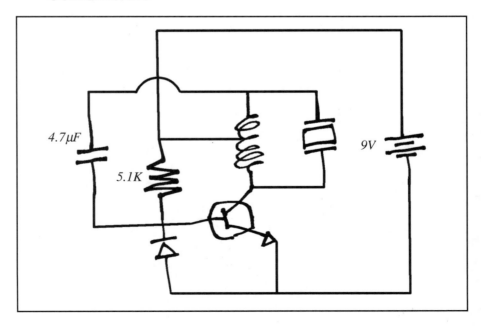

Instructions

Use the schematic illustration on the page at the left and the instructions below to build your pocket engine.

1. From the positive side of the battery, connect a wire to the 5.1KΩ ohm resistor.

2. Then connect a wire from the same side of the resistor to a lead of a 4.7µF capacitor.

3. From the other lead of the resistor, connect a wire to the center lead of the transistor.

4. From the center lead of the transistor, run a wire to the striped end of the diode.

5. From the other end of the diode, connect a wire to the negative side of the battery.

6. Then, from the left side (when viewed from the flat side) of the transistor, run a wire to the negative side of the battery.

7. From the right side of the transistor, connect a wire to the right lead of the transformer on the side that has three leads.

8. From the center lead of the three-lead side of the transformer, connect a lead to the positive side of the battery.

9. Connect a wire from the left lead of the three-lead side of the transformer to the negative side of the 4.7µF capacitor (the side with the black stripe.)

10. From the left side of the two-lead side of the transformer, connect a lead to the left side of the earphone.

Pocket Engine

11. From the right side of the two-lead side of the transformer, connect a lead to the right side of the earphone. Then, listen for a racecar engine sound.

How Come, Huh?

The noise you hear in your earphone should sound like the revving engine of a car. That's because the earphone, itself, actually makes the sound in much the same way an engine does.

Car engines produce their sound by creating a controlled explosion inside the engine. When the gas ignites, it expands, pushes the piston out of the way, and then escapes out the tailpipe. When the hot air exits the car, it expands, producing a pop. Firecrackers, jet engines, and bags full of air do the same thing.

A similar kind of thing is occurring inside your earphone. Inside is a small coil of copper and a tiny piece of plastic with a magnet glued to it. The circuit you built sends little zaps of electricity through the copper coil, which magnetizes it for just a second. Each time this happens, the magnet is attracted to the coil and pulls it over. After the electricity stops, the coil loses it magnetic field, and the magnet falls away. Because the coil is attached to the plastic, the plastic moves, and this pushes a little puff of air out of the earphone, causing a pop. When lots of these pops happen very quickly, it sounds like exhaust is coming out of the tailpipe of a car.

The next time you play a video game that makes a racecar sound, you hear the sound of a car driving by, or you hear a jet plane flying overhead, you'll know how the sound-effects were made.

Impromtu Cop Car

The Experiment

This circuit produces a sound similar to a police or ambulance siren, and it also has a flashing LED. This is not exactly the kind of stuff that you should use to make your mom pull over when she is hauling you to the skating rink, but after you build this, you will have an idea of how real emergency vehicles get your attention as they are driving down the road.

Materials

Resistor, 510Ω
Resistor, 5.1KΩ
Resistor, 10KΩ
Resistor, 100KΩ
Resistor, 470KΩ
Battery, 9V
Connecting wires,
    each 3 inches long
LED
Capacitor, .05μF
Transformer
2 Transistors
Piezo transducer
Yellow wire, 4 inches long

Schematic

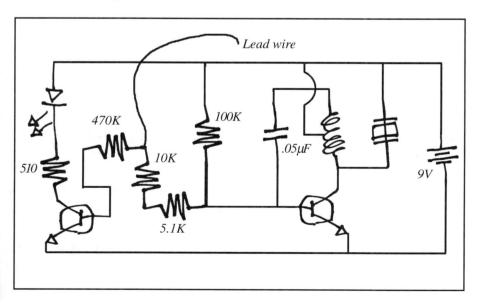

Impromtu Cop Car

Instructions

1. Connect a .05µF capacitor to the left side of the three-lead end of the transformer.

2. Connect the other lead of the .05µF capacitor to the lead of a 5.1KΩ ohm resistor.

3. From the same lead of the 5.1KΩ ohm resistor, run a wire to a lead of a 100KΩ ohm resistor.

4. From the same lead of the 100KΩ ohm resistor, run a wire to the center lead of a transistor.

5. From the right side of the transistor (when viewed from the flat side) connect a wire to the right lead of a transformer that has three leads.

6. Connect the lead of the transformer that is on the right when you look at it from the flat side to one lead of the piezo transducer.

7. Connect a lead of a 510Ω ohm resistor to the unused side of the 100KΩ ohm resistor.

8. From the same lead of the 100KΩ ohm resistor, connect a lead to the positive side of a battery.

9. From the positive side of the battery, run a wire to the left side, looking at it from the flat side, of a second transistor.

10. Coming off the same lead, run a yellow wire. Don't connect this to anything.

11. Looking at the first transistor from the flat side, connect a wire from the left-hand side of the first transistor to the right-hand side of the second transistor.

12. Looking at the second transistor from the flat side, connect a wire from the right-hand side of the second transistor to the negative side of the battery.

13. Connect a wire from the unused side of the 5.1KΩ ohm resistor to a 10KΩ ohm resistor.

14. From the other lead of the 10KΩ ohm resistor, connect a wire to a 470KΩ ohm resistor.

15. Connect the other lead of the 470KΩ ohm resistor to the center lead of the second transistor.

16. Connect the other lead of the 510Ω ohm resistor to the positive side of the LED.

17. Connect the left side of the second transistor to the negative side of the LED.

How Come, Huh?

This circuit incorporates something called an oscillator, which is a circuit that switches the electricity on and off very quickly—thousands of times every second. All of this electricity turns the piezo transducer on and off very quickly, causing it to vibrate very rapidly. That's what makes the sound. Here, you control the oscillator by touching the wires together, and this sends electricity to the oscillator and to the LED.

People can get pretty distracted when they drive cars, and that's why emergency vehicles, which need to pass all of these distracted drivers, use lights and sirens. The lights flash so rapidly and shine so brightly that they are almost impossible to ignore while you are driving down the same road. The quick-changing sound of a siren also gets people's attention immediately.

Theater Lighting

The Experiment
This circuit uses the stored electricity in a capacitor to slowly turn off the LED in your circuit. It gives you an idea of how lighting effects are created for movies and how light is controlled in buildings.

Materials
Resistor, 10KΩ
Resistor, 100KΩ
Resistor, 510Ω
Capacitor, 470µF
2 Transistors
LED
Battery, 9V
Yellow wire, 4 inches long
Connecting wires, each 3 inches long

Schematic

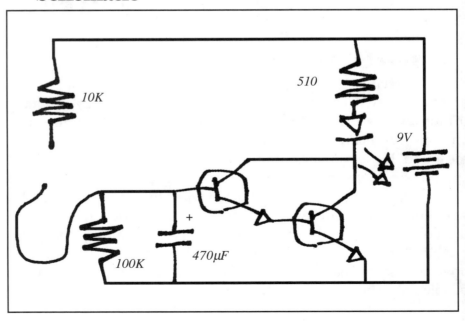

Instructions

1. From a 510Ω ohm resistor, connect a wire to a 10KΩ ohm resistor.

2. From the same side of the 10KΩ ohm resistor, connect a wire to the positive battery terminal.

3. From the other side of the 510Ω ohm resistor, connect a wire to the positive side of the LED.

4. Looking at a transistor from the flat side, connect the wire from the right to the right-hand wire of a second transistor.

5. Looking at the second transistor from the flat side, connect the right-hand wire to the negative side of the LED.

6. Looking at the first transistor from the flat side, connect the left-hand wire to the center of the second transistor.

7. From a lead of a 470µF capacitor, connect a wire to a 100KΩ ohm resistor.

8. From the same lead of the 100KΩ ohm resistor, connect a wire to the left side (when viewed from the flat side) of the second transistor.

9. Looking at the second transistor from the flat side, connect the left-hand wire to the negative side of the battery.

10. From the unconnected side of the 10KΩ ohm resistor, connect a wire to the yellow wire.

11. From the unused lead of the 470µF capacitor, connect a wire to the unused side of the 100KΩ ohm resistor.

Theater Lighting

12. From the same side of the 100KΩ ohm resistor, connect a wire to the center wire of the first transistor lead.

How Come, Huh?

This circuit uses a capacitor, which we talked about in the introduction. It's a storage tank that fills up with electricity so that you have a reserve to meet an increased demand at some time. In this project, the capacitor fills up with electricity and then passes it on to the part of the circuit where the LED is. When the electricity hits the LED, it initially has lots of punch and lights up the LED right away. As the current empties out of the capacitor, the light gets dimmer and dimmer, until it finally goes out.

Fading circuits are used in movie theaters to fade the lights slowly before the movie starts, which gives everyone a few moments to get ready before the room gets totally dark. It gives folks who are making their way toward their seats with a half-gallon of soda pop and a garbage bag full of popcorn a visual cue that they need to pack it in and get going.

It's also nice to have the lights slowly fade on at the end of the movie. After you've been sitting in the dark for about two hours, it is not a pleasant experience to suddenly have a bunch of bright lights shining in your face, causing stress to your pupils, flooding your cones and rods with light, and causing an unpleasant sensation.

Fading circuits are also used in video cameras to produce the fade-out and fade-in effects, in stage lights, and in the lights that are inside cars. Anytime you see a light that can be faded in or out, you know that it is running on stored energy from a capacitor.

Jingle Lights

The Experiment
When you hook up the battery to this circuit, the red and green LEDs will blink back and forth fairly rapidly. This is a great idea if you want to decorate for the holidays or if you are trying to build a traffic intersection for really fast drivers.

Materials

2 LEDs
Resistor, 100KΩ
Resistor, 470KΩ
Resistor, 510Ω
Connecting wires,
    each 3 inches long

2 Transistors
Battery, 9V
Capacitor, 4.7µF
Capacitor, 470µF

Schematic

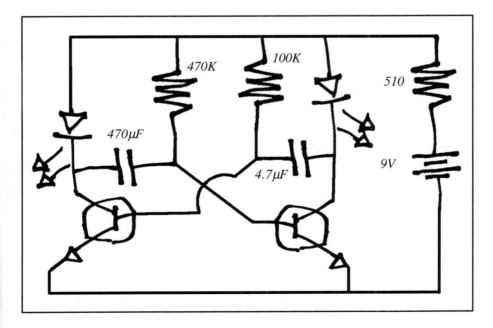

Jingle Lights

Instructions

1. From a 4.7µF capacitor, connect a wire to the negative side of the LED.

2. From the negative side of the LED, run another wire to the right side (when viewed from the flat side) of a transistor.

3. From a lead of a 510Ω ohm resistor, connect a wire to the 100KΩ ohm resistor.

4. From the same side of the 100KΩ ohm resistor, connect a wire to a 470KΩ ohm resistor.

5. From the same side of the 470KΩ ohm resistor, connect a wire to the positive side of the LED.

6. From the same, positive side of the LED, connect a wire to the positive side of a second LED.

7. From the unused side of the 470µF capacitor, connect a wire to the unused side of 100KΩ ohm resistor.

8. From the same lead of the 100KΩ ohm resistor, run a wire to the center lead of the transistor.

9. From the other side of the 470µF capacitor, connect a wire to the other side of the 470KΩ ohm resistor.

10. From the same lead of the 470KΩ ohm resistor, run a wire to the center lead of a second transistor.

11. From the other side of the 4.7µF capacitor, connect a wire to the right side (when viewed from the flat side) of the second transistor.

12. From the same lead on the right side of the transistor, connect a wire to the negative side of the second LED.

13. Looking at the first transistor from the flat side, connect the left-hand wire to the left side of the second transistor.

14. From the same lead on the left side of the transistor, connect a wire to the negative side of the battery.

15. From the unused side of the 510Ω ohm resistor, connect a wire to positive side of the battery.

How Come, Huh?

This circuit works like an old-fashioned traffic light, which had a built-in timer that controlled the color change. These timers were mechanical devices that used gears and switches to turn lights on and off to direct traffic. The timer in your project is a little more modern. It is made from capacitors and resistors.

When the capacitor fills with enough electrical current, it automatically changes which LED is lit. As one capacitor empties, the other is filling and getting ready to take over. The amount of time that each light stays lit is a function of the size of the capacitors and how much electricity they can hold.

Burglar Alarm

The Experiment

Before you connect the battery to this circuit, twist the ends of the two loose wires together, and then connect the battery. To activate the alarm, all you have to do is untwist the wires.

Materials

Resistor, 100KΩ
Resistor, 470KΩ
Resistor, 510Ω
Transformer
Yellow wire, 4 inches long
Connecting wires, each 3 inches long

Transistor
Capacitor, .01µF
Piezo transducer
Battery, 9V

Schematic

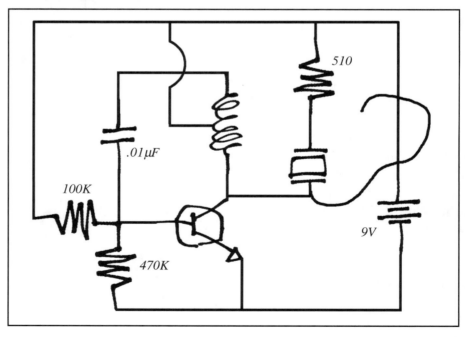

Instructions

1. Connect a wire from one lead of the .01µF capacitor to the left lead on the three-lead side of the transformer, looking at it from the three-lead side.

2. From a lead of a 510Ω ohm resistor, connect a wire to a lead of the piezo transducer.

3. From the same lead of the piezo transducer, connect the yellow wire.

4. From the other lead of the 510Ω ohm resistor, connect a wire to a lead of a 100KΩ ohm resistor.

5. From the same lead of the 100KΩ ohm resistor, connect a wire to the positive side of the battery.

6. From the positive side of the battery, connect a wire to the center lead of the transformer, on the side that has three leads coming out of it.

7. Looking at the transistor from the flat side, connect the left-hand wire to the right lead of the transformer that has three leads.

8. From the same lead of the transformer, connect a wire to the other lead of the piezo transducer.

9. From the same lead of the piezo transducer, connect a wire to the yellow wire.

10. From the unused lead of the .01µF capacitor, connect a wire to the unused lead of the 100KΩ ohm resistor.

Burglar Alarm

11. From the same lead of the 100KΩ ohm resistor, connect a wire to a 470KΩ ohm resistor.

12. From the same lead of the 470KΩ ohm resistor, connect a wire to the center lead of the transistor.

13. From the unused lead of the 470KΩ ohm resistor, connect a wire to left side (when viewed from the flat side) of the transistor.

14. From the same lead of the transistor, connect a wire to the negative side of the battery.

How Come, Huh?

This is one of the most common kinds of burglar alarm. Part of the circuit is a simple switch, with one part mounted on a door and the other on the door frame. If the door is closed, the switch is connected and the current can get through. When the door is opened, the switch opens, and the electricity can't get through, causing the piezo transducer to sound off in its typically annoying style.

The circuit is set up so that the current can either pass through the switch or go through the piezo transducer. Either way makes a complete circuit. Electrons in an electrical current are very practical. They always take the easiest path they can find. In this case, the part of the circuit with the piezo transducer has more resistance than the part with the switch. It's no contest—the current heads over the switch and leaves the piezo transducer alone. However, when someone opens the door and consequently the switch, the switch part of the circuit becomes a dead end. Suddenly, the piezo transducer is the only pathway available to the electrons, so they zip over there and the alarm buzzer sounds.

Some commercial burglar alarms work exactly the same way. Look for little wired boxes on store doors. Another version uses little strips of metal tape on the glass in store display windows. That metal tape is the "switch" part of the burglar alarm. If someone wants to rob the store and decides that the best way in is to break the glass, that breaks the metal tape, and the alarm sounds.

You can use this to wire the door to your room or the lid to your treasure box. If anyone opens the switch, the current continues to flow through the piezo transducer, and you will immediately know that you are being invaded.

Musical Line-Up

The Experiment

This circuit uses pencil lines to produce music. First, build the circuit, following the instructions we provide. Then find a piece of thin white paper and trace the funny shape on page 44. Using a regular (not colored) pencil, draw the shape so it's really dark and there is no white showing through at all. Press one wire to the bottom of the pattern and the other to the end of one of the three lines. Make sure your finger doesn't touch the metal ends of the wire when you press the wire onto the pencil marks. Try each line for a different musical note!

Materials

Resistor, 10KΩ
Resistor, 100KΩ
Transistor
Battery, 9V
Connecting wires, each 3 inches long
Capacitor, .01µF
Transformer
Piezo transducer
2 Yellow wires, each 4 inches long

Schematic

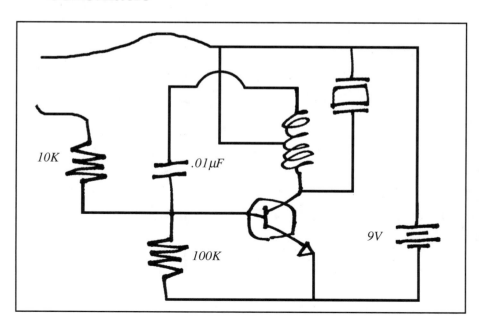

Instructions

1. Connect one end of a yellow wire to a lead of a 10KΩ ohm resistor.

2. Connect a different yellow wire to the positive side of the battery.

3. From the positive side of the battery, connect the wire to the center lead of a transformer, on the side that has three leads coming out of it.

4. From the same lead of the transformer, connect a wire to a lead of a piezo transducer.

5. Connect a wire from one lead of a .01µF capacitor to the unused lead of the 10KΩ ohm resistor.

6. From the same lead of the 10KΩ ohm resistor, connect a wire to a 100KΩ ohm resistor.

7. From the same lead of the 100KΩ ohm resistor, connect a wire to the center lead of a transistor.

8. From the unused lead of the .01µF capacitor, connect a wire to the left lead of the transformer, on the side that has three leads coming out of it.

9. From the right side of the transistor, looking at the transistor from the flat side, connect the right-hand wire to the right lead of the transformer, on the side that has three leads coming out of it.

10. From the same lead of the transformer, connect a wire to the other lead of the piezo transducer.

11. From the unused lead of the 100KΩ ohm resistor, connect

Musical Line-Up

a wire to the left side (when viewed from the flat side) of the transistor.

12. From the same lead of the transistor, connect a wire to the negative side of the battery.

How Come, Huh?

Despite the fact that we call the dark part of the pencil the "lead," there is actually no lead in a pencil at all. It's a material called graphite, which is an electrical conductor. When you put the wires at either end of the pencil line, electricity zips through the pencil line, making the oscillator oscillate and activating the piezo transducer.

The only problem is that graphite isn't a very good conductor. When the line you've drawn is really short, there is not very much resistance, and lots of electricity can get through. The longer the line is, the more resistance there is, and the harder it is for the electricity to make it from one wire to the other. The lines act kind of like a group of thin, flat resistors. That's why different lines produce different sounds.

All of the electronic instruments that musicians use in a concert utilize some variation on this circuit idea, whether they're keyboards, guitars, or whatever. The sound that is produced is a function of the amount of resistance that is provided by the circuit pathway.

Finger Switch

The Experiment

This circuit turns on an LED when you simply touch two wires to different parts of your same finger. This kind of circuit will show you how touch-screens are activated on computers, how elevators know how to respond to your touch, and how any other touch-activated switches work for your convenience.

Materials

Resistor, 510Ω
Resistor, 1KΩ
Resistor, 10KΩ
LED
2 Transistors
2 Yellow wires, each 4 inches long
Connecting wires, each 3 inches long
Battery, 9V

Schematic

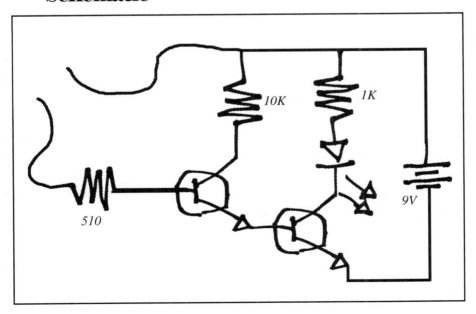

Finger Switch

Instructions

1. Connect one end of a yellow wire to a lead of a 510Ω ohm resistor.

2. Connect a different yellow wire to a 10KΩ ohm resistor.

3. From the same lead of the 10KΩ ohm resistor, connect a wire to a 1KΩ ohm resistor.

4. From the same lead of the 1KΩ ohm resistor, connect a wire to the positive side of the battery.

5. From the unused lead of the 10KΩ ohm resistor, connect a wire to the right side (when viewed from the flat side) of a transistor.

6. From the unused lead of the 510Ω ohm resistor, connect a wire to the center lead of the transistor.

7. Looking at the transistor from the flat side, connect the left-hand wire to the center of the second transistor.

8. From the unused lead of the 1KΩ ohm resistor, connect a wire to the positive lead of the LED.

9. From the right side of the second transistor, connect a wire to the negative side of the LED.

10. From the left side of the second transistor, connect a wire to the negative side of the battery.

How Come, Huh?

This kind of switch works because we are conductors, ourselves. Our bodies use electrical signals to do everything from pumping blood through our hearts, to squeezing air into and out of our lungs, to sending messages to our brains. So it makes sense that when electricity comes in contact with our skin, we become part of the circuit.

This switch has no moving parts, which makes it very useful and practical because there is nothing to wear out or break. That's why you'll find lots of elevators with touch-activated switches, because they get pushed hundreds of times a day. These switches also work really well in places where regular switches would be too big, such as in computer touch-screens. These screens are covered with lots of tiny little switches that turn on when you touch them, letting the computer know where your finger is. When your finger hits the screen, it connects two or more of the switches and sends a message to the computer that you have made the particular choice that you did. Technology has no bounds.

Auditory Circuit

The Experiment

This circuit uses sound waves to turn on the LED. After you connect the battery, clap right above the piezo transducer and you'll see the LED light up. It will flash for an instant when the sound waves from your clap hit the piezo transducer. Try shouting really loudly and see if you can light up the LED.

Materials

Resistor, 10KΩ
Resistor, 100KΩ
Resistor, 470KΩ
Resistor, 510Ω
Connecting wires, each 3 inches long

Piezo transducer
LED
2 Transistors
Battery, 9V

Schematic

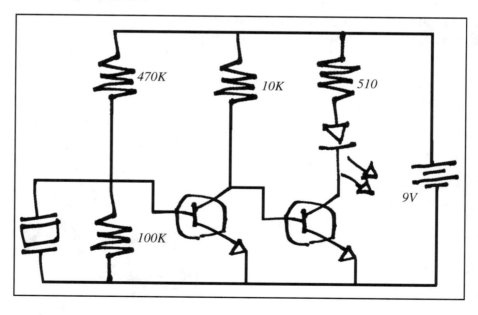

The Adventures of Diode Dude • B. K. Hixson

Instructions

1. Connect a wire from a 510Ω ohm resistor to a 10KΩ ohm resistor.

2. From the same lead of the 10KΩ ohm resistor, connect a wire to a 470KΩ ohm resistor.

3. From the same lead of the 10KΩ ohm resistor, connect a wire to the positive side of a battery.

4. From the unused lead of the 510Ω ohm resistor, connect a wire to the positive side of an LED.

5. Looking at the transistor from the flat side, connect the right-hand wire to the negative side of the LED.

6. From the unused lead of the 10KΩ ohm resistor, connect a wire to the center lead of the transistor.

7. From the same lead of the transistor, connect a wire to the right side (when viewed from the flat side) of a second transistor.

8. From a 100KΩ ohm resistor, connect a wire to the unused lead of the 470KΩ ohm resistor.

9. From the same lead of the 470KΩ ohm resistor, connect a wire to the center lead of the second transistor.

10. From the same lead of the transistor, connect a wire to a lead of a piezo transducer.

11. From the unused lead of the 100KΩ ohm resistor, connect a wire to the left side (when viewed from the flat side) of the first transistor.

Auditory Circuit

12. From the left lead of the first transistor, connect a wire to the left side of the second transistor.

13. Connect a wire from the second transistor to the negative side of the battery.

14. Connect a wire from the negative side of the battery to the unused lead of the piezo transducer.

How Come, Huh?

The circuit you have built was designed to respond to a loud, sharp sound. When you clap your hands, the air compresses, and it sends vibrations through the air. The vibrations are picked up by the disc inside the piezo transducer, which causes it to bend back and forth. These tiny vibrations induce an electrical current. (In this case, it's a *tiny* electrical current.) This little bit of electricity is just enough to start up the circuit and send enough electricity to the LEDs to get them to light up.

These kinds of circuits are very useful in situations where a small amount of sound is very unusual. Museums and bank vaults come to mind. Also, you have probably seen those TV ads where the person sitting in the chair claps her hands and the lights go off and on. It is the same kind of switch that you just created.

Photo Switch

The Experiment

This circuit uses light that is shining on a photocell to activate a circuit and light an LED. When you vary the amount of light that illuminates the photocell by going into a dark room, such as a closet or bathroom, the LED will light up. Turn on the lights, and the LED will dim and then eventually turn off. You can experiment with different levels of light to see how the LED responds by simply passing your hand over the photocell.

Materials

Photocell
Transistor
Battery, 9V
Resistor, 10KΩ
LED
Connecting wires,
    each 3 inches long

Schematic

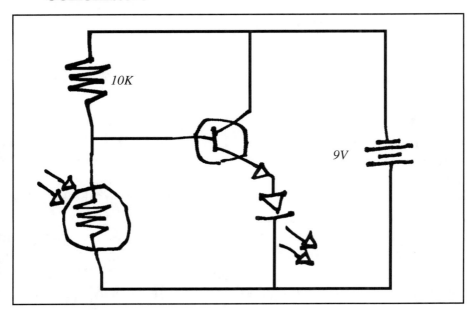

Photo Switch

Instructions

1. Connect a wire from a 10KΩ ohm resistor to the left-hand lead (when viewed from the flat side) of a transistor.

2. From the same lead of the transistor, connect a wire to the positive side of a battery.

3. From the unused lead of the 10KΩ ohm resistor, connect a wire to the center lead of the transistor.

4. From the same lead of the transistor, connect a wire to a lead of a photocell.

5. From the left lead of the transistor, connect a wire to the positive lead of an LED.

6. From the negative side of the LED, connect a wire to the negative side of the battery.

7. From the negative side of the battery, connect another wire to the unused lead of the photocell.

How Come, Huh?

The important part of this circuit is the photocell, or photoresistor. This was described in the introduction. The photoresistor controls the amount of electricity that can flow in response to light. In bright light, the photoresistor has low resistance, which means that electricity can easily flow through it. In darkness, it has high resistance, and little or no electricity can squeeze through the circuit. This circuit is designed to respond to the amount of electricity getting through the photocell. If the level of light reaching the photocell goes down too far, the LED responds by lighting up. When the room brightens again, the LED shuts off.

The most common use for this idea is in porch lights, nightlights, and street lights. Before we got a little more sophisticated with our electronics, somebody actually had to turn all the street lights on and off by hand. Eventually, the lights ran with mechanical timers, but the timers had to be readjusted for the changes in the length of day and Daylight Savings time.

These days, each light has its own light-sensitive circuit, so it can determine the level of light, regardless of the time of year, and switch itself on and off. If you have a nightlight, there's a good chance it also works with the same kind of circuit. To check and see if it does, indeed, have a photocell, cover it with your fingertip and see if you can change the resistance in the circuit and turn the light on.

Light Organ

The Experiment

This is an instrument that lets you play music with the aid of a little light. Once it's all hooked up, use this circuit in a well-lit room. Wave your hands a few inches over the photocell to change the tone you hear. The less light there is on the photocell, the lower the tone will be.

If you have trouble getting the tone to change, place an empty toilet paper tube directly over the photocell. This will restrict the source of light to only that light that comes directly from above. This way, when you wave your hand over the hole in the tube, you will have more control over the amount of light that hits the photocell.

Materials

Resistor, 100KΩ
Resistor, 470KΩ
Transistor
Piezo transducer
Connecting wires, each 3 inches long
Capacitor, .01µF
Photocell
Transformer
Battery, 9V

Schematic

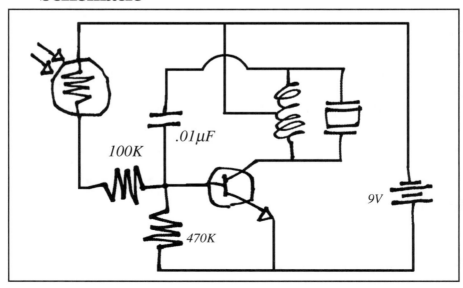

Instructions

1. From a .01.µF capacitor, connect a wire to the left lead of a transformer, from the side that has three leads.

2. From the same lead of the transformer, connect a wire to a lead of the piezo transducer.

3. From the .01.µF capacitor, connect a wire to a 100KΩ ohm resistor.

4. From the 100KΩ ohm resistor, connect a wire to a 470KΩ ohm resistor.

5. From the same lead of the 470KΩ ohm resistor, connect a wire to the center of a transistor.

6. Looking at the transistor from the flat side, connect the right-hand wire to the right lead of the transformer that has three leads.

7. From the same lead of the transformer, connect a wire to the unused lead of the piezo transducer.

8. From the positive battery lead, connect a wire to the center lead of the transformer.

9. From the same lead of the transformer, connect a wire to a photocell.

10. From the unused lead of the photocell, connect a wire to the unused lead of the 100KΩ ohm resistor.

11. From the unused lead of the 470KΩ ohm resistor, connect a wire to the left-hand lead (when viewed from the flat side) of the transistor.

Light Organ

12. From the same lead of the transistor, connect a wire to the negative side of the battery.

How Come, Huh?

You controlled sound in project #6 with pencil lines. This time, you're using light. The electrical current in this circuit, as with the previous project, is controlled by the amount of light passing through the photocell. The electricity goes into an oscillator. If there isn't much light, there isn't much electricity, and a low tone is produced. If there is a lot of light, there is a lot of electricity, and you get a higher pitch.

With a little practice, you can control the kinds of sounds your circuit makes. Move your hand around in different patterns. Try making a sound like a siren or spaceship, or use any other auditory distraction that you can concoct.

Photocells are used in lots of other ways, too. Cameras use them to keep pictures from getting too much or too little light. Some robots use photocells to find their way around by helping them sense where light and dark areas are in a room. Photocells have even been used in television sets to automatically adjust the brightness of the picture to match the lighting in the room.

Timing with Capacitance

The Experiment

In this experiment, after a battery is hooked up, the timer will start when you touch the loose end of the wire to the circuit. When you do this, the red LED will light up, will stay lit for roughly 20 seconds, and will then turn off by itself. It's normal if the LED still looks like it's slightly lit, even when it's off. Sometimes the timer will start itself when you connect the battery. If the LED is on when you connect the battery, wait about 20 seconds, and it should turn off.

Materials

Resistor, 1KΩ
Resistor, 5.1KΩ
Resistor, 10KΩ
Resistor, 100KΩ
Resistor, 470KΩ
Yellow wire,
    4 inches long

Capacitor, .01μF
Capacitor, 47μF
LED
2 Transistors
Battery, 9V
Connecting wires,
    each 3 inches long

Schematic

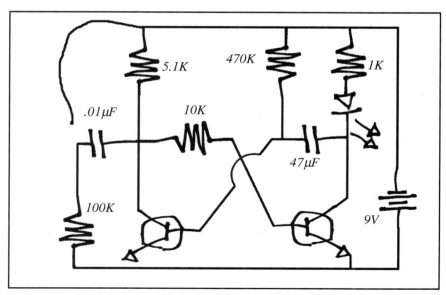

Timing with Capacitance

Instructions

1. From a .01µF capacitor, connect a wire to a 100KΩ ohm resistor.

2. From the unused lead of the .01µF capacitor, connect a wire to a 5.1KΩ ohm resistor.

3. From the same lead of the 5.1KΩ ohm resistor, connect a wire to a 10KΩ ohm resistor.

4. From the same lead of the 10KΩ ohm resistor, connect a wire to the right-hand lead (when viewed from the flat side) of a transistor.

5. From a 1KΩ ohm resistor, connect a wire to the unused lead of the 5.1KΩ ohm resistor.

6. From the same lead of the 5.1KΩ ohm resistor, connect a wire to a 470KΩ ohm resistor.

7. From the same lead of the 470KΩ ohm resistor, connect a wire to the positive side of a battery.

8. From the positive side of the battery, connect a wire.

9. From the unused lead of the 10KΩ ohm resistor, connect a wire to the center of the second transistor.

10. From a 47µF capacitor, connect a wire to the unused lead of the 470KΩ ohm resistor.

11. From the same lead of the 470KΩ ohm resistor, connect a wire to the center lead of the first transistor.

12. From the unused lead of the 47µF capacitor, connect a wire to right-hand lead (when viewed from the flat side) of the second transistor.

13. From the same lead of the second transistor, connect a wire to the negative side of an LED.

14. From the unused lead of the 100KΩ ohm resistor, connect a wire to the left-hand lead of the first transistor.

15. From the same lead of the first transistor, connect a wire to the left-hand lead (when viewed from the flat side) of the second transistor.

16. Looking at the second transistor from the flat side, connect the left-hand lead to the negative side of the battery.

17. From the 1KΩ ohm resistor, connect a wire to the positive side of the LED.

How Come, Huh?

This is another capacitor-based circuit. When you connect the wires, electricity flows into the capacitor and fills it, just like a little water tank. When the capacitor is full, it has enough power to switch on the part of the circuit that lights up.

The capacitor works like a storage tank that holds water until you need it. Then, when it lets it go, it dumps the whole thing, all at once. Assuming you were a vengeful and conniving kind of person, you would use this kind of water tank to soak your friend. You would get your friend positioned under the water, and then *whoosh*, away the water would go in one huge wave.

Timing with Capacitance

Capacitors work the same way. Even if you have only a small flow of electricity, you can use a capacitor to store the electricity up into a higher charge. The switch that turns on the light won't work until the capacitor has stored up enough electricity to make it go.

This is a very predictable capacitor that works like a timer. If you know how much electricity is going into the capacitor and how much charge it will hold, then you will be able to determine how long it will take for the capacitor to fill completely. By changing the resistors in the circuit, you can change how long it takes the timer to activate and deactivate.

By making circuits that are just a bit more complicated than this one, it's possible to build extremely accurate clocks. Many Olympic sports depend on clocks that measure speed in hundredths and thousandths of a second. It's not possible for a human timekeeper with a stopwatch to measure events with such precision, but these special clocks can do the job.

Piezo Candle

The Experiment

You can have fun with your friends, using this circuit. Tell them that you can extinguish the LED on your circuit, simply by blowing it out like a candle. They, of course, will think that you are goofier than a loon, but all you have to do is blow into the piezo transducer to prove them wrong.

Materials

Resistor, 1KΩ
Resistor, 5.1KΩ
Resistor, 10KΩ
Resistor, 510Ω
Connecting wires,
    each 3 inches long
2 Transistors
Piezo transducer
LED
Battery, 9V

Schematic

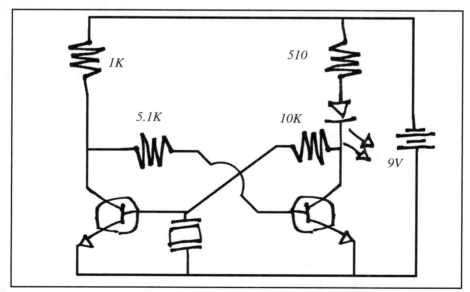

Piezo Candle

Instructions

1. From a 510Ω ohm resistor, connect a wire to a 1KΩ ohm resistor.

2. From the unused lead of the 1KΩ ohm resistor, connect a wire to the positive side of a battery.

3. From the unused lead of the 1KΩ ohm resistor, connect a wire to a 5.1KΩ ohm resistor.

4. From the same lead of the 5.1KΩ ohm resistor connect a wire to the right-hand lead (looking at it from the flat side) of a transistor.

5. From the unused lead of the 1KΩ ohm resistor, connect a wire to center of the second transistor.

6. From a 10KΩ ohm resistor, connect a wire to the center of the first transistor.

7. From the same lead of the first transistor, connect a wire to a lead of the piezo transducer.

8. From the unused lead of the 510Ω ohm resistor, connect a wire to the positive end of an LED.

9. From the unused lead of the 10KΩ ohm resistor, connect a wire to the right-hand lead (when viewed from the flat side) of the second transistor.

10. From the same lead of the second transistor, connect a wire to the negative side of the LED.

11. From the left-hand lead (when viewed from the flat side) of the second transistor, connect a wire to the right-hand lead of the first transistor.

12. Looking at the first transistor from the flat side, connect the left-hand lead to the negative side of the battery.

13. From the negative side of the battery, connect a wire to the unused lead of the piezo transducer.

How Come, Huh?

When you blow into the piezo transducer, your breath bends the metal disc inside, producing a small electrical current. Even though it's a very small amount of electricity, it's enough to trigger the "off" switch in your circuit, and your LED hits the sheets.

OK, so you can blow out an LED. Is there more to life than this...? Actually, there are other, more practical applications of this idea. One is weather. As the speed of the wind changes, the sound it makes blowing on a microphone changes. The microphone can be hooked up to a computer that will interpret the changes in sound and calculate the speed of the wind. This is almost as good as a wet finger.

You can also measure the amount of natural gas that comes into your home. There is a recently invented meter that has no moving parts and is the size of a small paperback book. If you have natural gas at your house, zip outside and look at how large your meter is. It's in the range of a mutant toaster. This new meter has a small microphone that "listens" to how fast the gas is coming through the meter. The small computer attached to the microphone can use this "speed" information to figure out how much gas is being used. The information is sent over the telephone line to the gas company. Then, they send you a bill. The joys of modern technology....

Conductivity Tester

The Experiment

This circuit allows you to test whether or not certain objects conduct electricity. After you hook up all the wires, just touch the ends of the two loose wires to whatever you want to test, and listen for a tone. If you hear a beep, you'll know that the thing you're testing is a conductor. If not, it's an insulator.

Materials

Resistor, 100KΩ
Transformer
Transistor
2 Yellow wires, each 4 inches long

Capacitor, .01µF
Piezo transducer
Battery, 9V
Connecting wires, each 3 inches long

Schematic

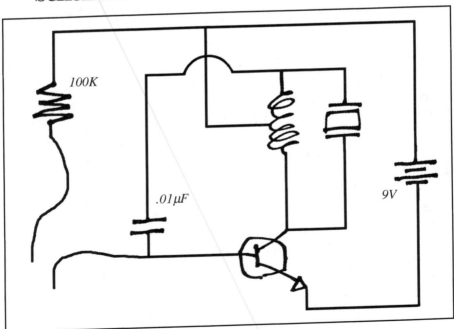

Instructions

1. Connect one end of a yellow wire to a lead of a 100KΩ ohm resistor.

2. Connect one end of a second yellow wire to a lead of a .01µF capacitor.

3. From the same lead of the .01µF capacitor, connect a wire to the center of a transistor.

4. From the unused lead of the 100KΩ ohm resistor, connect a wire to the positive side of a battery.

5. From the same lead on the positive side of the battery, connect a wire to the center lead of the transformer, on the side that has three leads coming out of it.

6. Looking at the transistor from the flat side, connect the right-hand lead of the transistor to the right-hand lead of the transformer, on the side that has three leads coming out of it.

7. From the same lead of the transformer, connect a wire to a lead of the piezo transducer.

8. From the unused lead of the .01µF capacitor, connect a wire to the left lead on the three-lead side of the transformer, looking at it from the three-lead side.

9. From the same lead of the transformer, connect a wire to the unused lead of the piezo transducer.

10. Looking at the transistor from the flat side, connect the left-hand wire to the negative side of the battery.

Conductivity Tester

How Come, Huh?

This circuit lets you test different objects to determine if they are conductors or insulators. When you touch the wires to an object that conducts electrons, the electricity zips through the circuit and produces a buzz. If the electricity can't go through, it doesn't make any sound.

Electricians use similar devices to check parts of things they're trying to fix, such as, say, an old television remote control. Sometimes a wire may be cut or broken, but may be inside or covered up where it can't easily be reached. An electrician could take the remote control completely apart, peel all of the covering off the wire, and see if it's okay. Using a circuit like yours, though, an electrician could simply touch the ends of the questionable wire to the conductivity tester. If the circuit makes a noise, the wire is fine. If the circuit is quiet, you need to replace the wire.

You can also use the circuit to test things in your house to see if they are conductors or insulators. Try the obvious items first, such as spoons and popsicle sticks. Then try some fun stuff. Try an apple, the hair on your pet gerbil, and the plastic trinket in your mom's purse. Be creative and have fun.

Electronic Rooster

The Experiment

This circuit activates the piezo transducer when light shines on it. After you connect the battery, you will hear the piezo transducer. Now cover the photocell with your hand. The piezo transducer should turn off. If it's still making noise, try taking the circuit into a dark room; if it shuts off, you know the circuit is working properly.

Materials

Resistor, 10KΩ
Resistor, 100KΩ
Resistor, 470KΩ
Transformer
Battery, 9V

2 Transistors
Capacitor, .01μF
Piezo transducer
Photocell
Connecting wires,
    each 3 inches long

Schematic

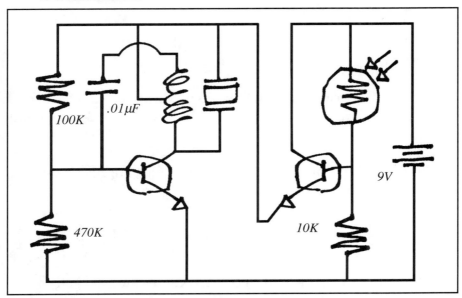

Electronic Rooster

Instructions

1. From a lead of a .01µF capacitor, connect a wire to a lead of a 100KΩ ohm resistor.

2. From the same lead of the resistor, connect a wire to a lead of a 470KΩ ohm resistor.

3. From the same lead of the 470KΩ ohm resistor, connect a wire to the center lead of a transistor.

4. From the unused side of the 100KΩ resistor, connect a wire to the left side of a second transistor, facing it from the flat side.

5. From the same lead of the transistor, connect another wire to the center lead of the 3-legged side of a transformer.

6. From the same lead of the transformer, connect a wire to a lead of a piezo transducer.

7. From the unused side of the .01µF capacitor, connect a wire to the left lead of the three-legged side of the transformer.

8. From the right side (when viewed from the flat side) of the first transistor, connect a wire to the right lead on the 3-legged side of the transformer.

9. From the same lead of the transformer, connect a wire to the unused lead of the piezo transducer.

10. From a 10KΩ resistor, connect a wire to a lead of the 470KΩ resistor.

11. From the same lead of the 470KΩ resistor, connect a wire to the left leg, facing the flat side, of the first transformer.

12. From the same leg of the transformer, connect a wire to the negative side of a battery.

13. From the right side of the second transistor (when viewed from the flat side), connect a wire to the positive side of the battery.

14. From the positive side of the battery, connect a wire to a lead of the photocell.

15. From the unused side of the 10KΩ resistor, connect a wire to the center lead of the second transistor.

16. From the same lead of the transistor, connect a wire to the unused lead of the photocell.

How Come, Huh?

The most important part of this circuit is the photocell. The more light that is available, the more electricity that is produced. We call this circuit the "Electronic Rooster" because you can put it in your bedroom window at night, and when the sun comes up, it will beep at you.

There are other fun things that you can do with your Electronic Rooster. You can wire the fridge to catch folks who raid the ice box at night, you can put it in your treasure drawer and as soon as anyone opens it up, it will buzz, and you can also use it as a burglar alarm if you keep cool stuff in your closet. Have fun and use your imagination.

Electron Hand Jive

The Experiment

Every kid has played the game of hand tag, where you hold your hands out flat, your partner puts his hands just over your hands, and the guy with the hands on the bottom of the sandwich tries to quickly slap one or both of the hands of the other person before that person can pull them away. This circuit will remove any doubt as to whether or not contact has been made.

Materials

Resistor, 100KΩ
Resistor, 470KΩ
Transistor
Battery, 9V
Connecting wires,
    each 3 inches long

Capacitor, .01µF
Transformer
Piezo transducer
2 Yellow wires,
    each 4 inches long

Schematic

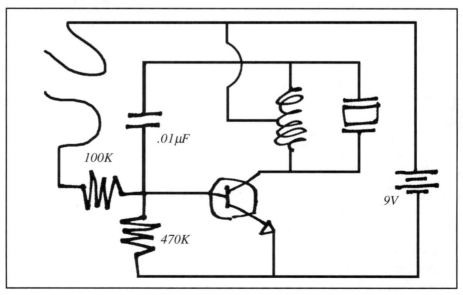

Instructions

1. From a .01µF capacitor, connect a wire to the left leg of the 3-legged side of a transformer.

2. From the same leg of the transformer, connect a wire to a lead of a piezo transducer.

3. From the positive side of a battery, connect a wire to the center leg on the 3-legged side of the transformer.

4. From the same leg of the transformer, connect a yellow wire.

5. Connect the yellow wire to a lead of a 100KΩ resistor.

6. From the unused side of the .01µF capacitor, connect a wire to the unused side of the 100KΩ resistor.

7. From the same lead of the 100KΩ resistor, connect a wire to a 470KΩ resistor.

8. From the same lead of the 470KΩ resistor, connect a wire to the center of a transistor.

9. From the right lead (facing the flat side of the transistor), connect a wire to the right leg of the 3-legged side of the transformer.

10. From the same leg of the transformer, connect a wire to the unused leg of the piezo transducer.

11. From the unused leg of the 470KΩ resistor, connect a wire to the left lead, facing the flat side of the transistor.

12. From the same leg of the transistor, connect a wire to the negative side of the battery.

Electron Hand Jive

How Come, Huh?

"I got you!" "Did not!" "Did too!" And so it goes. Everyone in North America has had this discussion. Technology to the rescue. You have now built a device, specially designed to solve this age-old dispute. The trick to the solution is that each person playing needs to hold onto a wire connected to your Electron Hand Jive circuit.

This circuit is so sensitive, it can tell when a little bit of electricity has passed through the bodies of two people holding wires connected to the circuit board. Find a partner and experiment with how well it works.

Now, on to the famous game that leads to all the discussion. Have two people hold one wire and stand facing each other. One person holds her hand palm-up, and the other person puts her hand palm-down, just over the top of the other hand, but without actually touching it. The person whose hand is on the bottom tries to slap the top of her partner's hand before she can pull it away.

You've probably played this game before, but now you have an electronic judge! Actually, the circuit is so sensitive that it will work through as many as six people! Form two chains of three people holding hands. Have one person from each group hold onto a wire. When the people at the other end of the chains touch, the piezo transducer should go off.

Morse's Code Machine

The Experiment

This circut makes a tone every time you touch the yellow wire to the battery.

Materials

Capacitor, .01µF
Resistor, 100KΩ
Transformer
Piezo transducer
Transistor
Battery, 9V
Yellow wire, 4 inches long
Connecting wires, each 3 inches long

Schematic

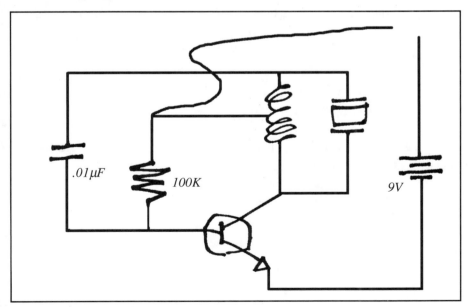

Morse's Code Machine

Instructions

1. From a lead of a .01µF capacitor, connect a wire to the left lead on the 3-legged side of a transformer.

2. From the same lead of the transformer, connect a wire to a lead of the piezo transducer.

3. From the unused lead of the .01µF capacitor, connect a wire to a lead of a 100KΩ resistor.

4. From the same lead of the resistor, connect a wire to the center lead of a transistor.

5. From the unused leg of the 100KΩ resistor, connect a wire to the center lead on the 3-legged side of the transformer.

6. From the right lead, facing the flat side, of the transistor, connect a lead to the right leg on the 3-legged side of the transformer.

7. From the same leg of the transformer, connect a wire to the unused leg of the piezo transducer.

8. From the left leg (viewed from the flat side) of the transistor, connect a wire to the negative side of a battery.

9. Connect a yellow wire to the positive side of the battery.

How Come, Huh?

When you touch the loose wire to the open connection, you complete the circuit and send electricity to the piezo transducer, which buzzes.

You now have an electronic version of a telegraph. The first telegraph in the United States was invented in 1837 by Samuel Morse, who also invented the code that was used to send information back and forth over the telegraph. This code is called, for obvious reasons, "Morse Code." It is still used today by shortwave radio operators to communicate all across the world.

When cables were strung along the railroad lines in the 1800s, spreading west across the United States, the telegraph became a viable method of communication. Messages could be sent instantly across distances that previously would have taken days or weeks by Pony Express or mail car. By replacing the speaker wire with longer pieces of wire from the hardware store, you can send messages from one room to another. Letters are spelled out by using "dots" (short sounds) and "dashes" (longer sounds).

To the right, you'll find the basic code.

Morse Code

A . -
B - . . .
C - . - .
D - . .
E .
F . . - .
G - - .
H . . . .
I . .
J . - - -
K - . -
L . - . .
M - -
N - .
O - - -
P . - - .
Q - - . -
R . - .
S . . .
T -
U . . -
V . . . -
W . - -
X - . . -
Y - . - -
Z - - . .

0 - - - - -
1 . - - - -
2 . . - - -
3 . . . - -
4 . . . . -
5 . . . . .
6 - . . . .
7 - - . . .
8 - - - . .
9 - - - - .

Simple Telephone

The Experiment

After you have completed this circuit, you will be able to have someone speak into the piezo transducer and you will be able to hear that person's voice. No long distance fees, no roaming, and unlimited minutes. Here is how you are going to build this most basic of electronic telephones.

Materials

Piezo transducer
Resistor, 470KΩ
Battery, 9V
Connecting wires,
    each 3 inches long

Capacitor, 470µF
Transistor
Earphone
Transformer

Schematic

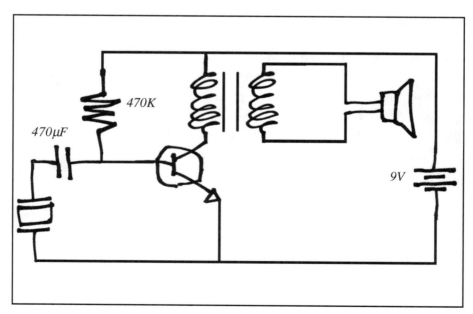

Instructions

1. From a lead of a 470µF capacitor, connect a wire to a lead of a piezo transducer.

2. From the left side, facing the flat side, of a transistor, connect a wire to the negative side of a battery.

3. From the battery, connect another wire to the unused leg of the piezo transducer.

4. From a leg of a 470KΩ resistor, connect a wire to the positive side of the battery.

5. From the positive side of the battery, connect a wire to the left leg on the 3-legged side of a transformer.

6. From the leg of the transistor that is on the right when you face its flat side, connect a wire to the right leg on the 3-legged side of the transformer.

7. From the unused lead of the 470µF capacitor, connect a wire to the unused lead of the 470KΩ resistor.

8. From the same lead of the resistor, connect a wire to the center lead of the transistor.

9. From a lead of an earphone, connect a wire to the right side, facing the 2-legged side, of the transformer.

10. Connect the unused lead of the earphone to the left leg on the 2-legged side of the transformer.

Simple Telephone

How Come, Huh?

The little disc that you talk into is actually a special kind of speaker, called a piezo (the term "piezo," appropriately, is Greek for "pressure") transducer. These piezo transducers are made of a crystalline material, and when this material is pushed by an air wave coming from your vocal cords, the pressure on the crystal produces electricity.

The electricity produced by the pressure travels down the wire to your circuit. It's a very small amount of electricity that, all by itself, won't do much, but your circuit uses a transistor to create an amplifier. The tiny bit of electricity going into the transistor creates a much stronger signal coming out—so strong, in fact, that it has enough energy to power the piezo transducer.

Piezo transducers and microphones are great because they have very few parts, they're simple to assemble, and they don't need much electricity to run. This makes them ideal for small, portable things, such as hand-held video games, electronic toys, cellular phone ringers, and things like beepers.

If you want to try a flavorful experiment, you can demonstrate the "piezo effect" with a Lifesaver. Take a Wintergreen Lifesaver into a totally dark room with a mirror. Pop the Lifesaver into your mouth and crush it between your molars, with your mouth open. You should see a spark where it collapses. The mechanical energy has been converted into electrical energy, exactly the same way that air pressure activates your piezo transducer.

Classic Crystal Radio

The Experiment

You are going to make a simple crystal radio. This is an age-old experiment that allows you to hear a very faint local radio station signal from the circuit that you are going to build. In addition to building the circuit, you will also want to amplify the sound coming from the piezo transducer. To do this, take a piece of paper and form a cone shape, with the opening of the cone being about the size of the transducer. Place the cone over the transducer to amplify the sound. You will also need to extend your antenna to strengthen your signal reception. Do this by unrolling the wire and stretching it across the room. Finally, you will need to ground your circuit. Run the loose wire to the metal leg of a table or chair and tape it in place. Once all of that is done, and you've also built the circuit, you can kick back and enjoy your tunes.

Materials

Transformer
Resistor, 100KΩ
Antenna
LED
Piezo transducer
Earphone
Diode
Capacitor, .001µF
Variable capacitor
Connecting wires, each 3 inches long

Schematic

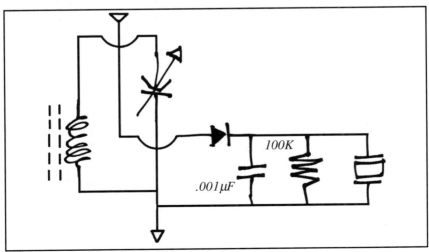

Classic Crystal Radio

Instructions

1. From the center leg of a variable capacitor, connect a wire to the lead of an antenna.

2. From a .001µF capacitor, connect a wire to a lead of a 100KΩ resistor.

3. From the same lead of the resistor, connect a wire to a lead of a piezo transducer.

4. From the same lead of the piezo transducer, connect a wire to the right leg of the variable capacitor.

5. From the same leg of the variable capacitor, connect a wire to another lead of the antenna.

6. From the lead of the antenna, connect a wire to the ground.

7. From the unused leg of the .001µF capacitor, connect a wire to the unused lead of the 100KΩ resistor.

8. From the same lead of the 100KΩ resistor, connect a wire to the unused lead of the piezo transducer.

9. From the same lead of the piezo transducer, connect a wire to the lead of the diode that has a stripe on it.

10. From the lead on the diode without the stripe, connect a wire to the center lead of the antenna.

How Come, Huh?

Your circuit is the simplest form of radio. Now that it is complete, you should be able to adjust the variable capacitor (by altering how much electricity it stores) and hear different radio stations through the earphone.

When a radio station sends out a signal, it sends out powerful radio waves from its antenna. Those waves travel through the air, like ripples traveling across a pond. Eventually, the radio waves hit something metal (such as your antenna) and they turn into electrical signals, which is why you don't need batteries—the radio waves create their own electricity.

This works only if your radio is grounded, so you have to attach one of the wires to a pipe, a piece of plumbing, or the leg of a table or chair. In reality, what you are doing is connecting Planet Earth into your circuit. Without grounding, the electricity from the radio waves has nowhere to go, and you won't be able to run the circuit, due to a huge blockage.

The electrical signals that come down your antenna into your circuit aren't much good at first. That's because your antenna is picking up lots of different radio signals. There are hundreds of different kinds of radio signals around us all the time, and if you could see radio waves, you wouldn't be able to see much of anything else—because they're absolutely everywhere. To be able to get something useful out of the antenna, the signals are filtered by a tuner, which allows only one frequency to pass through.

There's one more necessary step. The signal that made it by the tuner has the sound we want to listen to, but it's mixed in with the carrier wave. The last part of the circuit filters out the carrier wave and leaves only the electrical signals necessary to make your earphone work.

Transistor Radio

The Experiment

This radio uses a transistor to amplify the signal, so it's easy to hear from the piezo transducer. You will need to connect the antenna, which you should unravel completely and string all the way across a room. Make it as level as possible. Try this radio with and without the ground—the direct connection to the Earth. Turn the tuning knob to receive different stations.

Materials

Variable capacitor
Battery, 9V
Resistor, 1KΩ
Resistor, 100KΩ
Resistor, 470KΩ
Resistor, 510Ω
Connecting wires,
    each 3 inches long
Capacitor, .01µF
Capacitor, .001µF
Transistor
Piezo transducer
Diode
2 Transformers
Antenna

Schematic

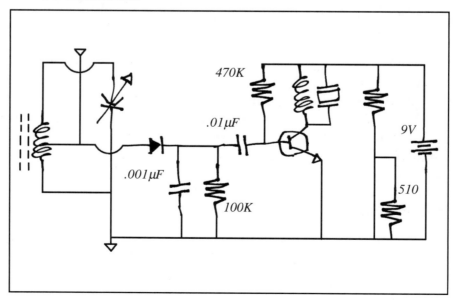

Instructions

1. From a 1KΩ resistor, connect a wire to a lead of a 470KΩ resistor.

2. From the same lead of the 470KΩ resistor, connect a wire to the positive side of a battery.

3. From the positive side of the battery, connect a wire to the lead that is on the left when you face the 3-legged side of a transformer.

4. From the same lead of the transformer, connect a wire to a lead of a piezo transducer.

5. From a lead of a 510Ω resistor, connect a wire to the negative side of the battery.

6. From the center lead of a variable capacitor, connect a wire to an end lead of the antenna.

7. From the side of the diode without a stripe, connect a lead to the center lead of the antenna.

8. From a lead of the .001µF capacitor, connect a wire to the unused lead of the 510Ω resistor.

9. From the unused lead of the 1KΩ resistor, connect a wire to a 100KΩ resistor.

10. From the same lead of the 100KΩ resistor, connect a wire to the lead of the transistor that is on the left when you face its flat side.

11. From the same lead of the transistor, connect a wire to the left lead of the variable capacitor.

Transistor Radio

    12. From the same lead of the variable capacitor, connect a wire the unused lead of the antenna.

    13. From the same lead of the antenna, connect a wire to the ground.

    14. From the unused lead of the .001µF capacitor, connect a wire to a lead of a .01µF capacitor.

    15. From the same lead of the .01µF capacitor, connect a wire to the unused lead of 100KΩ resistor.

    16. From the same lead of the 100KΩ resistor, connect a wire to the unused lead of the diode (the side with a stripe).

    17. From the unused lead of the .01µF capacitor, connect a wire to a lead of a 470KΩ resistor.

    18. From the same lead of the 470KΩ resistor, connect a wire to the center lead of the transistor.

    19. From the right-hand lead, facing the flat side of the transistor, connect a wire to the right-hand lead of a transformer, facing the 3-legged side.

    20. From the same lead of the transformer, connect a wire to the unused lead of the of the piezo transducer.

How Come, Huh?

The previous project used an old-fashioned crystal radio that caught a signal but really didn't have any way to amplify it and make it strong. That is why you made the cone and stuck it over the piezo transducer to help boost the signal. In this project, you should have noticed a distinct difference in the audibility of the signal—thanks to a little design called an amplifier.

An amplifier is a circuit that takes a weak electrical signal and makes it stronger. All an amplifier changes is the power of the signal; all the other information stays the same. It is kind of like giving a kid on a swing a push. The activity, direction, and excitement all stay the same; it is just amplitude of the swing that changes. You hear the same kind of thing with stereo advertisements. They claim to boost the signal with 100 kajillion watts per channel. The signal stays the same; it just gets a really big push.

Making the signal stronger has several important advantages. First, you can use the piezo transducer, which is larger than the earphone because there is enough electricity to power it. Second, you can listen to the radio without having to use a grounding wire, which makes it easier to carry around. Finally, you can hear weaker radio shows that you couldn't hear before with the earphone. Technology at its finest.

Flip-Flop Circuit

The Experiment

This circuit remembers which LED is lit. When you touch the wire to a certain spring, the circuit turns off one LED and turns on the other. Once the circuit is built, connect the battery, and the green LED will light up. Now touch the loose lead wire to the center lead of the first transistor. The red LED will light up, and the green LED will go off. Touch the lead wire to the center lead of the second transistor, and the green LED will light up again.

Materials

Resistor, 1KΩ
Resistor, 5.1KΩ
Resistor, 10KΩ
Resistor, 510Ω
Yellow wire, 4 inches long
2 LEDs
2 Transistors
Battery, 9V
Connecting wires,
    each 3 inches long

Schematic

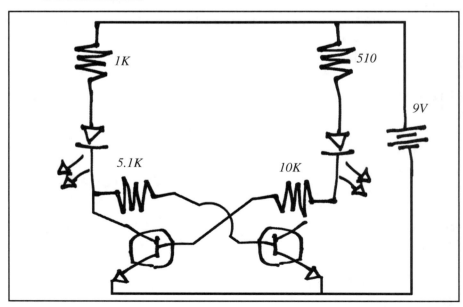

Instructions

1. From a lead of a 510Ω resistor, connect a wire to a lead of a 1KΩ resistor.

2. From the same lead of the 1KΩ resistor, connect a wire to the positive side of a battery.

3. From the unused lead of the 1KΩ resistor, connect a wire to the positive lead of an LED.

4. From a lead of a 5.1KΩ resistor, connect a wire to the right lead, facing the flat side of the transistor.

5. From the same lead of the transistor, connect a wire to negative side of the LED.

6. From a lead of a 10KΩ resistor, connect a wire to the center lead of the transistor.

7. From the unused lead of the 10KΩ resistor, connect a wire to the right lead, facing the flat side, of a second transistor.

8. From the same lead of the transistor, connect a wire to the negative side of a second LED.

9. From the unused lead of the 5.1KΩ resistor, connect a wire to the center lead of the second transistor.

10. From the left lead (viewed from the flat side) of the second transistor, connect a wire to the left lead, facing the flat side, of the first transistor.

11. From the same lead of the first transistor, connect a wire to the negative side of the battery.

Flip-Flop Circuit

12 From the negative side of the battery, connect a yellow wire.

13. From the unused lead of the 510Ω resistor, connect a wire to the positive side of the second LED.

How Come, Huh?

So what's a flip-flop? It is nothing less than the very heart and soul of computer logic and programming. Deep in the center of every computer, everything is based on this simple circuit. A flip-flop is a circuit that is changed from on to off and off to on with the same signal. It's like having a special light switch that works with a single button. The first time you push the button, the light goes on. The next time you push the button, the light goes off. One button does all the switching, which makes it easy to "flip" the light on and "flop" it back off again.

This is important because all computer information is stored as bits of information. A bit is defined as being either a one or a zero. These bits are recorded in tiny flip-flop circuits, where a flip-flop in the ON position means "one," and a flip-flop in the OFF position means "zero."

Computers need a generous supply of flip-flops. (Before technology allowed a lot of information to be stored in a very small space, the circuits in the first computers weighed several tons and required thousands of gallons of water to keep them cool when they were processing.) Information is translated into a code, almost like Morse code, that combines groups of ones and zeros to represent letters and numbers. To store one letter of information—a "q," for example—requires eight bits of information, shown as a pattern that looks something like 10011101. Each one or zero needs a flip-flop. The last paragraph would take 2,600 or so flip-flops! So you can see where more memory would be a very good thing. Go gigabytes!

Lots to think about. Hope you enjoyed the book. We will pick up with a new topic soon.

Until then, happy experimenting.

Science Fair Projects
•
A Step-by-Step Guide: From Idea To Presentation

Science Fair Projects

Ah, the impending science fair project—a good science fair project has the following five characteristics:

1. The student must come up with an *original* question.
2. That *original* question must be suited to an experiment in order to provide an answer.
3. The *original* idea is outlined with just one variable isolated.
4. The *original* experiment is performed and documented using the scientific method.
5. A presentation of the *original* idea in the form of a lab write-up and display board is completed.

Science Fair Projects

As simple as science report versus science fair project sounds, it gets screwed up millions of times a year by sweet, unsuspecting students who are counseled by sweet, unknowing, and probably just-as-confused parents.

To give you a sense of contrast, we have provided a list of legitimate science fair projects and then reports that do not qualify. We will also add some comments in italics that should help clarify why they do or do not qualify in the science fair project department.

Science Fair Projects
1. Temperature and the amount of time it takes mealworms to change to beetles.

Great start. We have chosen a single variable that is easy to measure: temperature. From this point forward, the student can read, explore, and formulate an original question that is the foundation for the project.

A colleague of mine actually did a similar type of experiment for his master's degree. His topic: The rate of development of fly larvae in cow poop as a function of temperature. No kidding. He found out that the warmer the temperature of the poop, the faster the larvae developed into flies.

2. The effect of different concentrations of soapy water on seed germination.

Again, wonderful. Measuring the concentration of soapy water. This leads naturally into original questions and a good project.

3. Crystal size and the amount of sugar in the solution.

This could lead into other factors, such as exploring the temperature of the solution, the size of the solution container, and other variables that may affect crystal growth. Opens a lot of doors.

vs. Science Reports

4. Helicopter rotor size and the speed at which the 'copter falls.

Size also means surface area, which is very easy to measure. The student who did this not only found the mathematical threshold with relationship to air friction, but she also had a ton of fun.

5. The ideal ratio of baking soda to vinegar to make a fire extinguisher.

Another great start. Easy to measure and track, and leads to a logical question that can either be supported or refuted with the data.

Each of these topics *measures* one thing, such as the amount of sugar, the concentration of soapy water, or the ideal size. If you start with an idea that allows you to measure something, then you can change it, ask questions, explore, and ultimately make a *prediction*, also called a *hypothesis*, and experiment to find out if you are correct. On the other hand, here are some well-meaning but misguided entries:

Science Reports, <u>not Projects</u>
1. Dinosaurs!
OK, great. Everyone loves dinosaurs, but where is the experiment? Did you find a new dinosaur? Is Jurassic Park alive and well, and are we headed there to breed, drug, or in some way test them? Probably not. This was a report on T. rex. Cool, but not a science fair project. And judging by the protest that this kid's mom put up when the kid didn't get his usual "A," it is a safe bet that she put a lot of time in and shared in the disappointment.

More Reports &

2. Our Friend the Sun

Another very large topic, no pun intended. This could be a great topic. Sunlight is fascinating. It can be split, polarized, reflected, refracted, measured, collected, and converted. However, this poor kid simply chose to write about the size of the sun, regurgitating facts about its features, cycles, and other astrofacts while simultaneously offending the American Melanoma Survivors Society. Just kidding about that last part.

3. Smokers' Poll

A lot of folks think that they are headed in the right direction here. Again, it depends on how the kid attacks the idea. Are they going to single out race? Heredity? Shoe size? What exactly are they after here? The young lady who did this report chose to make it more of a psychology-studies effort than a science project. She wanted to know family income, if smokers fought with their parents, how much stress was on the job, and so on. All legitimate concerns, but not placed in the right slot.

4. The Majestic Moose

If you went out and caught the moose, drugged it to see the side effects for disease control, or even mated it with an elk to determine if you could create an animal that would become the spokesanimal for the Alabama Dairy Farmers' Got Melk? promotion, that would be fine. But, another fact-filled report should be filed with the English teacher.

5. How Tadpoles Change into Frogs

Great start, but they forgot to finish the statement. We know how tadpoles change into frogs. What we don't know is how tadpoles change into frogs if they are in an altered environment, if they are hatched out of cycle, or if they are stuck under the tire of an off-road vehicle, blatantly driving through a protected wetland area. That's what we want to know—how tadpoles change into frogs, if, when, or under what measurable circumstances.

Now that we have beaten the chicken squat out of this introduction, we are going to show you how to pick a topic that can be adapted to become a successful science fair project. But first ... one more thought.

One Final Comment

A Gentle Reminder

Quite often, I discuss the scientific method with moms, dads, teachers, and kids, and get the impression that, according to their understanding, there is one, and only one, scientific method. This is not necessarily true. There are lots of ways to investigate the world we live in and on.

Paleontologists dig up dead animals and plants but have no way to conduct experiments on them. They're dead. Albert Einstein, the most famous scientist of the last century and probably on everybody's starting five of all time, never did experiments. He was a theoretical physicist, which means that he came up with a hypothesis, skipped over collecting materials for things like black holes and space-time continuums, and didn't experiment on anything or even collect data. He just went straight from hypothesis to conclusion, and he's still considered part of the scientific community. You'll probably follow the six steps we outline, but keep an open mind.

Project Planner

This outline is designed to give you a specific timeline to follow as you develop your science fair project. Most teachers will give you 8 to 11 weeks notice for this kind of assignment. We are going to operate from the shorter timeline with our suggested schedule, which means that the first thing you need to do is get a calendar.

A. The suggested time to be devoted to each item is listed in parentheses next to that item. Enter the date of the Science Fair and then, using the calendar, work backward, entering dates.

B. As you complete each item, enter the date that you completed it in the column between the goal (due date) and project item.

Goal   Completed   Project Item

1. Generate a Hypothesis (2 weeks)

Goal	Completed	Project Item
_____	_____	Review Ideas Contained in Labs
_____	_____	Try Several Experiments
_____	_____	Hypothesis Generated
_____	_____	Finished Hypothesis Submitted
_____	_____	Hypothesis Approved

2. Gather Background Information (1 week)

Goal	Completed	Project Item
_____	_____	Concepts/Discoveries Written Up
_____	_____	Vocabulary/Glossary Completed
_____	_____	Famous Scientists in Field

& Timeline

Goal   *Completed*   *Project Item*

3. Design an Experiment (1 week)

Goal	Completed	
_____	_____	Procedure Written
_____	_____	Lab Safety Review Completed
_____	_____	Procedure Approved
_____	_____	Data Tables Prepared
_____	_____	Materials List Completed
_____	_____	Materials Acquired

4. Perform the Experiment (2 weeks)

_____	_____	Scheduled Lab Time

5. Collect and Record Experimental Data (part of 4)

_____	_____	Data Tables Completed
_____	_____	Graphs Completed
_____	_____	Other Data Collected and Prepared

6. Present Your Findings (2 weeks)

_____	_____	Rough Draft of Paper Completed
_____	_____	Proofreading Completed
_____	_____	Final Report Completed
_____	_____	Display Completed
_____	_____	Oral Report Outlined on Index Cards
_____	_____	Practice Presentation of Oral Report
_____	_____	Oral Report Presentation
_____	_____	Science Fair Setup
_____	_____	Show Time!

Scientific Method
• Step 1 •
The Hypothesis

The Hypothesis

A hypothesis is an educated guess. It is a statement of what you think will probably happen. It is also the most important part of your science fair project because it directs the entire process. It determines what you study, the materials you will need, and how the experiment will be designed, carried out, and evaluated. Needless to say, you need to put some thought into this part.

There are four steps to generating a hypothesis:

Step One • Pick a Topic
This should be something that you are interested in studying. We would like to politely recommend that you take a peek at physical science ideas (physics and chemistry) if you are a rookie and this is one of your first shots at a science fair project. These kinds of lab ideas allow you to repeat experiments quickly. There is a lot of data that can be collected, and there is a huge variety to choose from.

If you are having trouble finding an idea, all you have to do is pick up a compilation of science activities (like this one) and start thumbing through it. Go to the local library or head to a bookstore and you will find a wide and ever-changing selection to choose from. Find a topic that interests you and start reading. At some point, an idea will catch your eye, and you will be off to the races.

Develop an Original Idea

Step Two • Do the Lab

Choose a lab activity that looks interesting, and try the experiment. Some kids make the mistake of thinking that all you have to do is find a lab in a book, repeat the lab, and you are on the gravy train with biscuit wheels. Your goal is to ask an ORIGINAL question, not repeat an experiment that has been done a bazillion times before.

As you do the lab, be thinking not only about the data you are collecting, but of ways you could adapt or change the experiment to find out new information. The point of the science fair project is to have you become an actual scientist and contribute a little bit of new knowledge to the world.

You know that they don't pay all of those engineers good money to sit around and repeat other people's lab work. The company wants new ideas, so if you are able to generate and explore new ideas, you become very valuable, not only to the company but to society. It is the question-askers that find cures for diseases, create new materials, figure out ways to make existing machines energy-efficient, and change the way that we live. For the purpose of illustration, we are going to take a lab titled, "Prisms, Water Prisms," from another book, *Photon U*, and run it through the rest of the process. The lab uses a tub of water, an ordinary mirror, and light to create a prism that splits the light into the spectrum of the rainbow. Cool. Easy to do. Not expensive and open to all kinds of adaptations, including the four that we discuss on the next page.

Step Three • Bend, Fold, Spindle, & Mutilate Your Lab
Once you have picked out an experiment, ask if it is possible to do any of the following things to modify it into an original experiment. You want to try to change the experiment to make it more interesting and find out one new, small piece of information.

Heat it	Freeze it	Reverse it	Double it
Bend it	Invert it	Poison it	Dehydrate it
Drown it	Stretch it	Fold it	Ignite it
Split it	Irradiate it	Oxidize it	Reduce it
Chill it	Speed it up	Color it	Grease it
Expand it	Substitute it	Remove it	Slow it down

If you take a look at our examples, that's exactly what we did to the main idea. We took the list of 24 different things that you could do to an experiment—not nearly all of them, by the way—and tried a couple of them out on the prism setup.

Double it: Get a second prism and see if you can continue to separate the colors further by lining up a second prism in the rainbow of the first.

Reduce it: Figure out a way to gather up the colors that have been produced and mix them back together to produce white light again.

Reverse it: Experiment with moving the flashlight and paper closer to the mirror and farther away. Draw a picture and be able to predict what happens to the size and clarity of the rainbow image.

Substitute it: You can also create a rainbow on a sunny day using a garden hose with a fine-spray nozzle attached. Set the nozzle adjustment so that a fine mist is produced and move the mist around in the sunshine until you see the rainbow. This works better if the sun is lower in the sky; late afternoon is best.

Hypothesis Worksheet
Step Three (Expanded) • Bend, Fold, Spindle Worksheet

This worksheet will give you an opportunity to work through the process of creating an original idea.

A. Write down the lab idea that you want to mangle.

_____
_____
_____
_____

B. List the possible variables you could change in the lab.
   i. _____
   ii. _____
   iii. _____
   iv. _____
   v. _____

C'MON. HE SAID TO STRETCH IT.

C. Take one variable listed in section B and apply one of the 24 changes listed below to it. Write that change down and state your new lab idea in the space below. Do that with three more changes.

Heat it	Freeze it	Reverse it	Double it
Bend it	Invert it	Poison it	Dehydrate it
Drown it	Stretch it	Fold it	Ignite it
Split it	Irradiate it	Oxidize it	Reduce it
Chill it	Speed it up	Color it	Grease it
Expand it	Substitute it	Remove it	Slow it down

   i. _____
_____
_____
_____

ii. _____
_____
_____

iii. _____
_____
_____

iv. _____
_____
_____

Step Four • Create an Original Idea—Your Hypothesis
Your hypothesis should be stated as an opinion. You've done the basic experiment, you've made observations, you're not stupid. Put two and two together and make a PREDICTION. Be sure that you are experimenting with just a single variable.

A. State your hypothesis in the space below. List the variable.
i. _____
_____
_____

ii. Variable Tested: _____

Sample Hypothesis Worksheet

On the previous two pages is a worksheet that will help you develop your thoughts and a hypothesis. Here is sample of the finished product to help you understand how to use it.

A. Write down the lab idea that you want to mutilate.
A mirror is placed in a tub of water. A beam of light is focused through the water onto the mirror, producing a rainbow on the wall.

B. List the possible variables you could change in the lab.
  i. **Source of light**
  ii. **The liquid in the tub**
  iii. **The distance from flashlight to mirror**

C. Take one variable listed in section B and apply one of the 24 changes to it. Write that change down and state your new lab idea in the space below.

The shape of the beam of light can be controlled by making and placing cardboard filters over the end of the flashlight. Various shapes, such as circles, squares, and slits will produce rainbows of different qualities.

D. State your hypothesis in the space below. List the variable. Be sure that when you write the hypothesis, you are stating an idea and <u>not asking a question.</u>

Hypothesis: The narrower the beam of light, the tighter, brighter, and more focused the reflected rainbow will appear.

Variable Tested: The opening on the filter

Scientific Method
• Step 2 •
Gather Information

Gather Information

Read about your topic and find out what we already know. Check books, videos, the Internet, and movies, talk with experts in the field, and grab an encyclopedia or two. Gather as much information as you can before you begin planning your experiment.

In particular, there are several things that you will want to pay special attention to and that should accompany any good science fair project.

A. Major Scientific Concepts

Be sure that you research and explain the main idea(s) that is / are driving your experiment. It may be a law of physics, a chemical rule, or an explanation of an aspect of plant physiology.

B. Scientific Words

As you use scientific terms in your paper, you should also define them in the margins of the paper or in a glossary at the end of the report. You cannot assume that everyone knows about geothermal energy transmutation in sulfur-loving bacteria. Be prepared to define some new terms for them ... and scrub your hands really well when you are done, if that is your project.

C. Historical Perspective

When did we first learn about this idea, and who is responsible for getting us this far? You need to give a historical perspective with names, dates, countries, awards, and other recognition.

Building a Research Foundation

1. This sheet is designed to help you organize your thoughts and give you some ideas on where to look for information on your topic. When you prepare your lab report, you will want to include the background information outlined below.

   A. Major Scientific Concepts (Two is plenty.)
      i. _____
   _____
   _____
   _____

      ii. _____
   _____
   _____
   _____

   B. Scientific Words (No more than 10)
      i. _____
      ii. _____
      iii. _____
      iv. _____
      v. _____
      vi. _____
      vii. _____
      viii. _____
      ix. _____
      x. _____

   C. Historical Perspective
      Add this as you find it.

2. There are several sources of information that are available to help you fill in the details from the previous page.

   A. Contemporary Print Resources
      (Magazines, Newspapers, Journals)
         i. _____
        ii. _____
       iii. _____
        iv. _____
         v. _____
        vi. _____

   B. Other Print Resources
      (Books, Encyclopedias, Dictionaries, Textbooks)
         i. _____
        ii. _____
       iii. _____
        iv. _____
         v. _____
        vi. _____

   C. Celluloid Resources
      (Films, Filmstrips, Videos)
         i. _____
        ii. _____
       iii. _____
        iv. _____
         v. _____
        vi. _____

D. Electronic Resources
   (Internet Website Addresses, DVDs, MP3s)
   i. _____
   ii. _____
   iii. _____
   iv. _____
   v. _____
   vi. _____
   vii. _____
   viii. _____
   ix. _____
   x. _____

E. Human Resources
   (Scientists, Engineers, Professionals, Professors, Teachers)
   i. _____
   ii. _____
   iii. _____
   iv. _____
   v. _____
   vi. _____

You may want to keep a record of all of your research and add it to the back of the report as an Appendix. Some teachers who are into volume think this is really cool. Others, like myself, find it a pain in the tuchus. No matter what you do, be sure to keep an accurate record of where you find data. If you quote from a report word for word, be sure to give proper credit with either a footnote or parenthetical reference. This is very important for credibility and accuracy. This will keep you out of trouble with plagiarism (copying without giving credit).

Scientific Method
• Step 3 •
Design Your Experiment

Acquire Your Lab Materials

The purpose of this section is to help you plan your experiment. You'll make a map of where you are going, how you want to get there, and what you will take along.

List the materials you will need to complete your experiment in the table below. Be sure to list multiples if you will need more than one item. Many science materials double as household items in their spare time. Check around the house before you buy anything from a science supply company or hardware store.

Material	Qty.	Source	$
1.			
2.			
3.			
4.			
5.			
6.			
7.			
8.			
9.			
10.			
11.			
12.			

Total $_____

Outline Your Experiment

This sheet is designed to help you outline your experiment. If you need more space, make a copy of this page to finish your outline. When you are done with this sheet, review it with an adult, make any necessary changes, review safety concerns on the next page, prepare your data tables, gather your equipment, and start to experiment.

In the spaces below, list what you are going to do in the order in which you are going to do it.

i. _____

_____

_____

ii. _____

_____

_____

iii. _____

_____

_____

iv. _____

_____

_____

v. _____

_____

_____

Evaluate Safety Concerns

We have included an overall safety section in the front of this book on pages 14–16, but there are some very specific questions you need to ask, and prepare for, depending on the needs of your experiment. If you find that you need to prepare for any of these safety concerns, place a check mark next to the letter.

____ A. *Goggles & Eyewash Station*
If you are mixing chemicals or working with materials that might splinter or produce flying objects, goggles and an eyewash station or sink with running water should be available.

____ B. *Ventilation*
If you are mixing chemicals that could produce fire, smoke, fumes, or obnoxious odors, you will need to use a vented hood or go outside and perform the experiment in the fresh air.

____ C. *Fire Blanket or Fire Extinguisher*
If you are working with potentially combustible chemicals or electricity, a fire blanket and extinguisher nearby are a must.

____ D. *Chemical Disposal*
If your experiment produces a poisonous chemical or there are chemical-filled tissues (as in dissected animals), you may need to make arrangements to dispose of the by-products from your lab.

____ E. *Electricity*
If you are working with materials and developing an idea that uses electricity, make sure that the wires are in good repair, that the electrical demand does not exceed the capacity of the supply, and that your work area is grounded.

____ F. *Emergency Phone Numbers*
Look up and record the following phone numbers for the Fire Department: _____ , Poison Control: _____ , and Hospital: _____ . Post them in an easy-to-find location.

Prepare Data Tables

Finally, you will want to prepare your data tables and have them ready to go before you start your experiment. Each data table should be easy to understand and easy for you to use.

A good data table has a **title** that describes the information being collected, and it identifies the **variable** and the **unit** being collected on each data line. The variable is *what* you are measuring, and the unit is *how* you are measuring it. They are usually written like this:

Variable (unit), or to give you some examples:

Time (seconds)
Distance (meters)
Electricity (volts)

An example of a well-prepared data table looks like the sample below. We've cut the data table into thirds because the book is too small to display the whole line.

Determining the Boiling Point of Compound X_1

Time (min.)	0	1	2	3	4	5	6
Temp. (°C)							

Time (min.)	7	8	9	10	11	12	13
Temp. (°C)							

Time (min.)	14	15	16	17	18	19	20
Temp. (°C)							

Scientific Method
• Step 4 •
Conduct the Experiment

Lab Time

It's time to get going. You've generated a hypothesis, collected the materials, written out the procedure, checked the safety issues, and prepared your data tables. Fire it up. Here's the short list of things to remember as you experiment.

_____ A. *Follow the Procedure and Record Any Changes*

Follow your own directions specifically as you wrote them. If you find the need to change the procedure once you are into the experiment, that's fine; it's part of the process. Be sure to keep detailed records of the changes. When you repeat the experiment a second or third time, follow the new directions exactly.

_____ B. *Observe Safety Rules*

It's easier to complete the lab activity if you are in the lab rather than in the emergency room.

_____ C. *Record Data Immediately*

Collect temperatures, distances, voltages, revolutions, and any other variables, and immediately record them into your data table. Do not think you will be able to remember them and fill everything in after the lab is completed.

_____ D. *Repeat the Experiment Several Times*

The more data that you collect, the better. It will give you a larger database, and your averages will be more meaningful. As you do multiple experiments, be sure to identify each data set by date and time so that you can separate them out.

_____ E. *Prepare for Extended Experiments*

Some experiments require days or weeks to complete, particularly those with plants and animals or the growing of crystals. Prepare a safe place for your materials so your experiment can continue undisturbed while you collect the data. Be sure you've allowed enough time for your due date.

Scientific Method
• Step 5 •
Collect and Display Data

Types of Graphs

This section will give you some ideas on how you can display the information you are going to collect as a graph. A graph is simply a picture of the data that you gathered, portrayed in a manner that is quick and easy to reference. There are four kinds of graphs described on the next two pages. If you find you need a leg up in the graphing department, we have a book in the series that will guide you through the process.

Line and Bar Graphs

These are the most common kinds of graphs. The most consistent variable is plotted on the "x," or horizontal, axis and the more temperamental variable is plotted along the "y," or vertical, axis. Each data point on a line graph is recorded as a dot on the graph, and then all of the dots are connected to form a picture of the data. A bar graph starts on the horizontal axis and moves up to the data line.

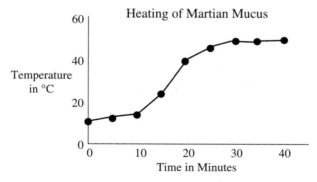

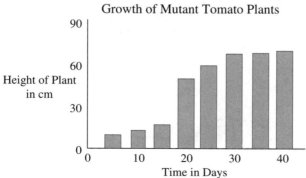

Best Fit Graphs
A best fit graph was created to show averages or trends rather than specific data points. The data that has been collected is plotted on a best fit graph, just as on a line graph, but instead of drawing a line from point to point to point, which sometimes is impossible anyway, you just freehand a line that hits "most of the data."

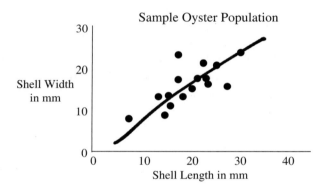

Pie Graphs
Pie graphs are used to show relationships between different groups. All of the data is totaled up, and a percentage is determined for each group. The pie is then divided to show the relationship of one group to another.

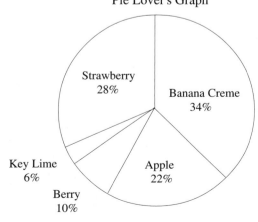

Other Kinds of Data

1. Written Notes & Observations

This is the age-old technique used by all scientists. Record your observations in a lab book. Written notes can be made quickly as the experiment is proceeding, and they can then be expounded upon later. Quite often, notes made in the heat of an experiment are revisited during the evaluation portion of the process, and they can shed valuable light on how or why the experiment went the way it did.

2. Drawings

Quick sketches as well as fully developed drawings can be used as a way to report data for a science experiment. Be sure to title each drawing and, if possible, label what it is that you are looking at. Drawings that are actual size are best.

3. Photographs, Videotapes, and Audiotapes

There are usually better than drawings, quicker, and more accurate, but you do have the added expense and time of developing the film. However, they can often capture images and details that are not usually seen by the naked eye.

4. The Experiment Itself

Some of the best data you can collect and present is the actual experiment itself. Nothing will speak more effectively for you than the plants you grew, the specimens you collected, or that big pile of tissue that was an armadillo you peeled from the tread of an 18-wheeler.

Scientific Method
• Step 6 •
Present Your Ideas

Oral Report Checklist

It is entirely possible that you will be asked to make an oral presentation to your classmates. This will give you an opportunity to explain what you did and how you did it. Quite often, this presentation is part of your overall score, so if you do well, it will enhance your chances for one of the bigger awards.

To prepare for your oral report, your science fair presentation should include the following components:

Physical Display
- _____ a. freestanding display board
    - hypothesis
    - data tables, graphs, photos, etc.
    - abstract (short summary)
- _____ b. actual lab setup (equipment)

Oral Report
- _____ a. hypothesis or question
- _____ b. background information
    - concepts
    - word definitions
    - history or scientists
- _____ c. experimental procedure
- _____ d. data collected
    - data tables
    - graphs
    - photos or drawings
- _____ e. conclusions and findings
- _____ f. ask for questions

Set the display board next to you on the table. Transfer the essential information to index cards. Use the index cards for reference, but do not read from them. Speak in a clear voice, hold your head up, and make eye contact with your peers. Ask if there are any questions before you finish and sit down.

Written Report Checklist

Next up is the written report, also called your lab write-up. After you compile or sort the data you have collected during the experiment and evaluate the results, you will be able to come to a conclusion about your hypothesis. Remember, disproving an idea is as valuable as proving it.

This sheet is designed to help you write up your science fair project and present your data in an organized manner. This is a final checklist for you.

To prepare your write-up, your science fair report should include the following components:

    _____ a.   binder
    _____ b.   cover page, title, & your name
    _____ c.   abstract (one paragraph summary)
    _____ d.   table of contents with page numbers
    _____ e.   hypothesis or question
    _____ f.   background information
                        concepts
                        word definitions
                        history or scientists
    _____ g.   list of materials used
    _____ h.   experimental procedure
                        written description
                        photo or drawing of setup
    _____ i.   data collected
                        data tables
                        graphs
                        photos or drawings
    _____ j.   conclusions and findings
    _____ k.   glossary of terms
    _____ l.   references

Display Checklist

Prepare your display to accompany the report. A good display should include the following:

Freestanding Display
- _____ a.   freestanding cardboard back
- _____ b.   title of experiment
- _____ c.   your name
- _____ d.   hypothesis
- _____ e.   findings of the experiment
- _____ f.   photos or illustrations of equipment
- _____ g.   data tables or graphs

Additional Display Items
- _____ h.   a copy of the write-up
- _____ i.   actual lab equipment setup

Index

Index

Antenna, 22
Auditory circuit, 48–50

Battery, 23
Bread board, 24
Building a research foundation, 107–109

Capacitor, 17
Conductivity, 64–66
Crystal radio, 79–81

Data, prepare tables for, 114
Diode, 19

Earphone, 22

Flip-flop circuit, 86–88

Graphs
    bar, 118
    best fit, 119
    line, 118
    pie, 119
    types of, 118–119

Home-school parent, 11–12
How to use this book, 8–13
Hypothesis
    definition, 99
    sample, 104
    worksheet, 102–103